Fostering Productivity and Competitiveness in Agriculture

OECD

Please cite this publication as:
OECD (2011), *Fostering Productivity and Competitiveness in Agriculture*, OECD Publishing.
http://dx.doi.org/10.1787/9789264166820-en

ISBN 978-92-64-16680-6 (print)
ISBN 978-92-64-16682-0 (PDF)

Photo credits: Cover © mipan - Fotolia.com.

Corrigenda to OECD publications may be found on line at: *www.oecd.org/publishing/corrigenda*.

Foreword

The growing global demand for food, feed and bio-fuel is well established, with population growth, and even more importantly income growth, increasing the quantity and changing the composition of agricultural commodity demand. Increasing output by bringing more land into production and using more water is technically feasible, but there are competing uses for these finite resources; water use is particularly constrained, with some forecasts suggesting that more food will need to be produced in the future with significantly less water than used today. And of course the impacts of climate change, while highly uncertain, likely imply important changes in water availability and perhaps even shifts in production zones.

The globally shared challenge, then, is to ensure greater efficiency in the use of available land and water resources; improving agricultural productivity is an essential requirement to increasing global food supplies on a sustainable basis.

Available measures of agricultural productivity growth – total factor productivity, land and labour productivity, and commodity yields – reveal a complex picture over time and across countries. In brief, productivity growth rates appear to be slowing, most notably in more developed countries where productivity levels are currently highest; productivity growth in recent years in particular has been higher in many less developed countries.

In light of the decline in real prices for agricultural commodities over the past 100 years, these trends may not be surprising. But higher prices since the early 2000s, and the prospects of higher prices over the coming decade, are changing the incentives for increased investment in agricultural productivity growth. The global food and agriculture system has consistently out-performed Malthusian expectations and can be expected to continue to do so. If current disincentives to an effective supply response by competitive suppliers are removed, and positive incentives put in their place to unleash innovation in the sector, it will contribute much to improving income, employment, and growth prospects, and thereby reducing poverty in many developing countries, and to building global food security.

A renewed focus on defining concrete actions to improve agricultural productivity growth on a sustainable basis is needed, now. Three broad areas requiring attention seem clear: closing the gap between actual and potential productivity levels in developing country agriculture; investing in agricultural innovation, broadly defined; and, improving national and international research collaboration.

OECD has launched work in each of these areas, and anticipates collaborating close with other international organisations and national governments, with a view to providing timely policy insights for use at national, regional and international levels.

This report is based on a document declassified by the OECD Working Party on Agricultural Policies and Markets in September 2011.

Table of contents

Tables

Figures

Executive Summary

In a context of limited resources and high input costs, ensuring stronger productivity growth in the agri-food sector is essential to successfully respond to increased and more diversified demands for food and non-food use of agricultural products. Productivity growth has been the subject of renewed attention as recent developments in agricultural markets have reinforced concerns on global food security, sustainability, and on the challenges resulting from climate change. Productivity growth also reflects relative developments in agricultural competitiveness across firms, sectors and countries. Depending on the viewpoint, competitiveness can be defined as the ability to face competition successfully, to sell products that meet demand requirements and, at the same time, ensure profits over time, or the aptitude to gain market shares, and most people agree that competitiveness is a relative concept which should be measured according to a benchmark. While productivity is generally considered as an important indicator of competitiveness, there are other measures including trade-related indicators, and strategic management measures such as firm-level indicators of production costs or profitability. Ideally a combination of indicators would provide a better picture of competitiveness at the farm, sector or national levels.

Productivity is an absolute concept and represents the ability to turn production inputs into outputs which can be measured at the farm, industry or national level. While total factor productivity can be used to measure the efficiency of all inputs being converted into all outputs, there are also partial indicators of productivity such as output per worker, per hectare (e.g. crop yield) or per animal (e.g. milk yield). Total factor productivity growth can be decomposed into three elements: 1) technological change, which indicates a change in the technology available (innovation creation); 2) technical efficiency, which represents the ability of farms to use best technologies available, and 3) scale efficiency. These components of total factor productivity are often used to measure innovation, creation and diffusion. It is, however, also possible to measure the adoption of a specific form of innovation.

Agricultural productivity developments depend on the period covered, the sector and the country considered, and on the indicator chosen. In developed countries, total factor productivity grew strongly from the 1960s to the mid-1990s, but not thereafter. Some studies indicate that productivity growth has slowed since the mid-1990s (e.g. Alston *et al.*, 2010). Over the same time period, the situation is diverse in other countries. On the one hand, agricultural productivity growth resumed in some transition economies after a temporary slowdown in the 1990s and is high in some large producing countries of Central and Eastern Europe. Moreover, productivity growth has been particularly strong in some emerging economies like Brazil and China. On the other hand, agricultural productivity growth is still low in most least-developed countries. Overall, there is no clear evidence that total factor productivity growth is decreasing at global level (Alston *et al.*, 2010, Chapter 4).

Among the possible determinants of productivity growth, research and development (R&D) has been the subject of many studies and receives specific attention in this report. R&D is the main source of new technologies and agricultural productivity growth in the long run. Agricultural R&D activities take place in private, public and farmers' organisations, as well as on-farm. Expenditure on R&D is often used as an indicator of efforts in this area, while the number of patents is considered as a measure of achievements. There are many conceptual models of how R&D leads to innovation and there have been many attempts to measure the impact of R&D expenditures on productivity growth in agriculture.

OECD's R&D database shows that the public sector is usually the main actor in providing agricultural R&D investment. The public expenditure on agricultural R&D has dominated the total agricultural R&D expenditure and has increased over the last decades in many OECD countries both in real terms and as a share of agricultural GDP. The private expenditure on agricultural R&D is also increasing in some countries, compensating in some cases for a decline in public R&D expenditure. OECD's Producer Support Estimate (PSE) database shows the share of R&D investment in total support for agriculture has increased in many countries over the past few decades. In addition, more diverse sources of agricultural R&D funding have emerged in recent years with the agricultural industry funding sometimes R&D activities undertaken by public institutions, while governments funded agricultural R&D activities undertaken by the private sector in other cases.

There is a wide diversity in the share of expenditures on agricultural R&D undertaken by public institutions as a percentage of agricultural GDP (R&D intensity) across OECD countries: in 2006 it ranged from less than 1% to more than 4%. In developing countries, agricultural R&D is undertaken almost entirely by public institutions and at a lower level of intensity than in OECD countries, usually below 1% of agricultural GDP.

Estimating the impact of agricultural R&D on productivity growth is challenging due to a number of attribution problems and lack of reliable data. First, there are many factors other than organised agricultural R&D that can affect productivity growth in agriculture. Second, there are spill-over effects across industries, regions and countries. Agriculture could be a major beneficiary of R&D undertaken in other areas of science or industries. The spatial spill-over of R&D benefits may also be significant. Third, research takes a long time to affect production, and then affects production for a long time. Estimating the impact of R&D on productivity therefore depends crucially on the research lag structure that is assumed. The estimated benefits of agricultural R&D generally far exceed its costs, with the literature reporting annual internal rates of return that range between 20% and 80%. According to some authors, the benefits of agricultural R&D are often under-estimated and this may result in under-investment (Alston, 2010).

Among other potential determinants of productivity and competitiveness, farm size has been the subject of numerous investigations, particularly in the context of structural change. A wide range of results is found depending on the circumstances and the type of indicators of size and competitiveness chosen. Conflicting results are also found regarding the relationship between technical efficiency and the share of hired labour and rented land, the time spent off the farm, or the age of the farm manager. But education always has a positive impact on farm performance. Differences in competitiveness across farms also depend on the natural environment in which they operate, including climatic conditions, soil quality and slope. While public policy is expected to affect farm competitiveness, its impact depends on the type of measure (Latruffe, 2010).

The review of the literature and existing data presented in this report identifies several gaps in the methodology and data. Further analysis with more sophisticated methodology and data is required to improve the understanding of the relationship between agricultural and agri-food productivity growth, competitiveness, and R&D, which plays a crucial role by introducing new technology and knowledge. The literature suggests there are many other factors that determine productivity growth, or more broadly the competitiveness of this sector. This suggests that a more comprehensive analytical framework going beyond the linear relationship between R&D expenditure and productivity growth would need to be adopted in future work to analyse "innovation systems" in agriculture. This more systemic approach suggests that innovation policy goes far beyond research expenditures and involves a wide range of public and private institutions that can affect incentives, knowledge sharing and the processes used for commercialisation. The analysis of existing database on R&D expenditures indicates the emergence of diverse patterns for agricultural R&D in terms of source of funding and institutional arrangements. Future work would take a closer look at institutional arrangements in agricultural innovation and knowledge systems, and examine the respective roles for the government versus the private sector in strengthening innovation systems and facilitating technological adoption, including research collaboration across sectors; protection of intellectual property rights; and knowledge flow. It is also suggested that a comprehensive effort be undertaken to measure different stages of the innovation system, including technological adoption and diffusion at the farm level, and to investigate the impact of agricultural policies on technological change and technical efficiency.

Chapter 1

Context and issues

This chapter outlines the importance of productivity growth for the sustained competitiveness of the agri-food sector and the role of agricultural research and development in fostering the innovations conducive to higher productivity growth. It briefly describes the content of the report and raises the main questions it attempts to respond to.

Background

Competitiveness is essential for economic growth. Productivity growth is a major element of sustained competitiveness and is largely linked to the adoption of new technologies or other innovations. In turn, innovation is driven by research and development (R&D) and influenced by other public policies. The strength of these theoretical relationships is a topical question for applied research. In the food and agriculture sector, fostering innovation to increase productivity growth, to ensure sustainable resource use, and to respond to demands from consumers is high on the national, regional and global policy agendas.

This report is the outcome of an OECD project on factors that determine innovation and competitiveness in agriculture In this context, two reports reviewing the literature on farm productivity and competitiveness (Latruffe, 2010) and on the impact of R&D on productivity growth (Alston, 2010) were prepared by consultants and are published as *OECD Working Papers on Food, Agriculture and Fisheries* No. 30 and 31 respectively.[1]

This consolidated report reviews and discusses current knowledge on the linkages between agricultural competitiveness, productivity growth, innovation, R&D and public policy. It draws on Latruffe (2010) and selected other recent sources (e.g. Alston *et al.*, 2010) to review evidence on productivity and competitiveness, and to discuss factors determining productivity growth, including the role of policies. Alston (2010) is drawn heavily upon to describe what is known about the measurement of productivity growth and relationship between R&D and productivity growth in agriculture. Finally, as a means to help identify what data is available the report summarises the agricultural research expenditure data available in OECD databases on agricultural support[2] and on research and development[3], and in the Agricultural Science and Technology Indicators (ASTI) database.[4]

Issues

While national competitiveness is often the focus of policy, Michael Porter argues that this focus is misplaced and should be directed toward productivity. "*Productivity depends both on the value of a nation's products and services, measured by the prices they can command in open markets, and the efficiency with which they can be produced. Productivity growth supports high wages, a strong currency, and attractive returns to capital — and with them a high standard of living. Productivity is the goal, not exports per se or whether firms operating in the country are domestic or foreign owned.*" (Porter *et al.*, 2007, p. 52).

If rival firms are improving their productivity and altering the terms of trade, a firm must become more productive to maintain its competitive position. This is especially true in agriculture where global innovations continue to reduce real commodity prices. Domestic and regional governments are increasingly aware of this and are looking for policy instruments to enhance productivity growth.

According to economic theory, productivity growth in the long run requires innovation, which can be defined as "the introduction of new or significantly improved goods or services, or the use of new inputs, processes, organisational or marketing methods (OECD and Eurostat Oslo Manual, 2005). To become more productive firms must be able to change their production systems over time. These innovations can be as simple as changing crops that are produced, or more complex, for example developing a new business model with entirely different production technologies. Economies of scale are also a component of productivity growth for individual firms (Latruffe, 2010).

While the ability to innovate and become more productive is determined in part within the firm (or the farm), it can also be affected by the economic and policy environment the firm is operating within (Porter *et al.*, 2007). This environment may depend on competitors, upstream firms supplying inputs and knowledge, and/or downstream firms involved in product marketing and utilisation. Governments have recognised that much of a firms' ability to innovate can be driven by public R&D, infrastructure, regulations, taxation, and other public policies that have both direct and indirect effects on the operating environment of firms.

Private firms are often unable to capture the full value from knowledge creation and knowledge transmission, in part because of the long lag between initial investments in agricultural R&D and the impact on farms' total factor productivity performance. This has limited the private incentive to invest in agricultural R&D. Historically, many OECD countries addressed this market failure by direct government involvement in agricultural R&D (e.g. the US land grant system; the National Institute for Agricultural Research in France; Agriculture and Agri-Food Canada in Canada) and some emerging economies, such as Brazil, China or South Africa, have also invested significantly. Over time, agricultural innovations have diffused within and between countries, with significant spill-over effects from developed and emerging economies to developing countries. Recognising the need for international collaboration, a number of international or regional networks on agri-food innovation systems have been developed, such as the Consultative Group on International Agricultural Research (CGIAR).

The global agricultural research landscape has changed a great deal over the last 20 years. With the advent of biotechnology and stronger intellectual property rights large multinational firms have made significant investment in crop research.[5] At the same time, public funding for agricultural research has increased more slowly or decreased in many countries while the research mandate has been broadened to include environmental, food and other issues. The recent increase in commodity prices, concerns of a possible slowing down of total factor productivity growth, and growing fiscal pressures, have led policy makers to become increasingly interested in the effectiveness of R&D as a means to increase productivity at the firm, industry and national level for competitiveness, and to respond more generally to global food security, sustainability and climate change challenges.[6] This has led many countries to review the functioning of their agricultural knowledge systems, to develop better evaluation indicators, and to explore ways to facilitate the creation, diffusion and adoption of innovations at the national, regional and global levels.

Coverage and content

The general questions addressed in the report are:

- What do we currently know about the conceptual linkages between agricultural competitiveness, productivity growth and innovation?
- What do we currently know about recent trends in agricultural productivity growth in OECD countries, emerging economies and around the globe?
- What data exists within the OECD and ASTI databases regarding agricultural R&D expenditures?
- What does the literature say about the determinants of productivity growth and competitiveness in OECD countries and emerging economies?
- What is the empirical evidence of linkages between R&D and agricultural productivity growth?
- Where are the gaps in data and the analysis required for a comprehensive understanding of how R&D affects agricultural productivity growth?

The report is organised as follows. Chapter 2 briefly reviews the economic definitions of agricultural competitiveness, productivity growth, innovation and R&D and then discusses how these relate to one another. Chapter 3 provides evidence on developments in agricultural productivity and competitiveness across countries on the basis of published material. Chapter 4 describes trends in agricultural R&D expenditures using OECD and ASTI data and identifies the gaps in the data available. Chapter 5 discusses the factors that explain productivity growth and technical efficiency, with a focus on the empirical evidence of linkages between agricultural R&D and productivity growth. Chapter 6 provides final remarks and suggests potential areas of further study.

Notes

1. Available at: *www.oecd.org/document/25/0,3746,en_2649_37401_47887193_1_1_1_37401,00.html.*

2. Information on the OECD database on Producer Support Estimates (PSE) and other agricultural support indicators can be found at: *www.oecd.org/agriculture/pse*.

3. OECD statistics on research and innovation are available at: *www.oecd-ilibrary.org/science-and-technology/data/oecd-science-technology-and-r-d-statistics_strd-data-en*.

4. Information on the Agricultural Science and Technology Indicators (ASTI) database can be found at: *www.asti.cgiar.org*.

5. Some evidence of increases in yields and margins per hectare from use of genetically modified crops was presented at the OECD Conference on Agricultural Knowledge Systems, held in Paris on 15-17 June 2011 (see presentations at: *www.oecd.org/document/20/0,3746,en_2649_37401_47217428_1_1_1_37401,00.html*).

6. This was the focus of the OECD Conference on Agricultural Knowledge Systems: Responding to Global Food Security and Climate Change Challenges, organised in Paris, on 15-27 June 2011 (*www.oecd.org/document/20/0,3746,en_2649_37401_47217428_1_1_1_37401,00.html*).

Chapter 2

Understanding agricultural competitiveness and productivity

This chapter provides definitions for the concepts used in the report: competitiveness, productivity, total factor productivity and its components, efficiency and innovation. It explains how they can be measured and how they relate to each other. In the report, productivity is considered as an indicator of competitiveness.

Competitiveness

Competitiveness is a relative concept that has most value when used at the firm level. Latruffe (2010) defines competitiveness as "the ability to face competition and to be successful when facing competition." or "the ability to sell products that meet demand requirements (price, quality, quantity) and, at the same time, ensure profits over time that enables the firm to thrive."

The ability of firms to compete or to be competitive depends on their operating economic environment. Superior technology (e.g. disease resistant varieties), local resource endowments (e.g. land, human capital), infrastructure (e.g. transportation and communications), and supportive institutions (e.g. product grading, auction markets) can increase profitability by increasing output prices, lower input costs and increase the efficiency of production.

Governments can also choose to foster a particular industry through policies that enhance profitability by using subsidies or regulatory powers to raise the output price, reduce input costs, or more generally through some macro-economic policies. As indicated by welfare economics, these policies, while benefitting a particular sector, come at a larger overall cost to the economy that results in lower economic surplus for the economy.

Krugman (1996, 2001) argues that the "obsession" with the export competitiveness of a country "is not only wrong but is also dangerous". He argues that competitiveness measured by notions such as the trade surplus of a country tend to favour protectionist policies over those that will enhance domestic and international economic growth, and can incite trade or currency wars. Porter and Krugman observe that "true competitiveness is measured by productivity" and "productivity is the goal, not exports *per se*". This provides a compelling case for investing resources to improve understanding of the relationship between R&D and productivity growth.

In an extensive review of the literature, Latruffe (2010) provides a critical review of various types of competitiveness indicators, discusses measurement issues and explains how they influence the findings on competitiveness. These indicators measure competitiveness from two perspectives including the measurement of strategic management such as production costs, profitability, productivity and efficiency, and the measurement of trade-related competitiveness such as revealed comparative advantage summarised in Tables of Annex A of this report. Latruffe (2010) concludes that competitiveness is "a broad, multifaceted, concept with no general agreement on how to define it, or how to measure it precisely," and that it is a relative measure used to compare firms and sectors within a domestic economy or firms and sectors within international markets. She notes that many authors estimate several indicators of competitiveness in parallel and suggests that it would be better to measure several components and aggregate them into a single measure of competitiveness in order to get a complete view.

Measuring productivity and its components

Firms use technology to combine inputs to produce output for the purpose of maximising profit. Generally speaking, productivity represents a firm's ability to convert production inputs into production outputs. A more "productive" firm has a higher ratio of output to input than a less productive firm. Productivity growth refers to the change in output/input ratios over time (Figure 2.1).

Figure 2.1. Graphical illustration of productivity growth

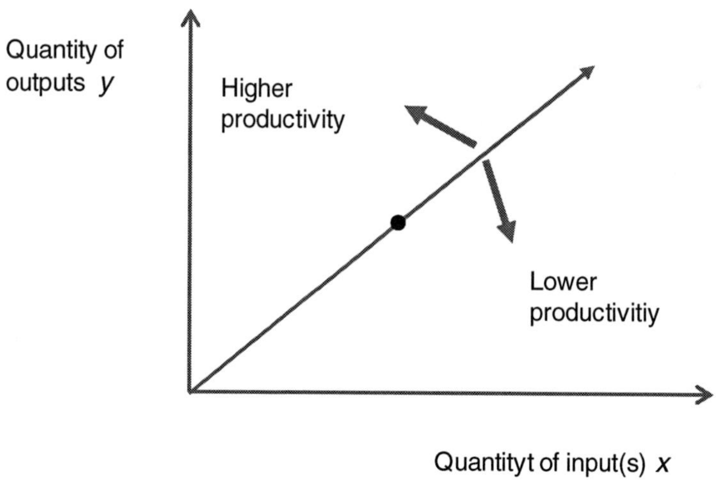

For measures of partial factor productivity, an index of output over a particular input is used to measure how output per unit of a particular input changes over time. Output per worker is a measure of labour productivity and yield per hectare is often used to describe land productivity. While partial factor productivity measures are useful for some purposes, such of examining labour markets or land markets, they can be misleading indicators of technological progress because they do not reflect changes in the use of other inputs. For example, a programme that heavily subsidises fertilisers would increase both land and labour productivity, but reduces overall (total factor) productivity and economic surplus.

Total Factor Productivity (TFP) can be defined as an index of total outputs over an index of total inputs. As such, TFP is a single measure designed to capture how efficiently a firm uses total inputs to produce outputs. Since TFP indices are sensitive to the way that various outputs and various inputs are aggregated, different aggregation approaches may lead to different estimation, with each consistent with a specific assumption on the underlying production function. As described by Latruffe (2010), the main TFP indices used in the literature include Laspeyres, Paasche, Fisher, Törnqvist-Theil and Eltetö-Köves-Szulc (EKS) indices.

While earlier studies employed the Laspeyre and Paasche indexes, these methods have largely been replaced by the Törnqvist-Theil index, which is consistent with a more flexible translog production function, and by the Fischer index, which is proved by Diewert (1992) to be a superlative index. The EKS method has also been employed in the estimation process by imposing transitivity to ensure comparability of estimation results between countries, regions or firms. There has also been a move to use discrete versions of the Divisia Index, which uses updated rather than constant weights to construct the input and output indexes.

Where firm-level data are available it is possible to construct an efficiency frontier, either econometrically or non-parametrically using linear programming methods referred to as Data Envelopment Analysis (DEA). The efficiency frontier represents the most productive input-output combinations observed in the industry. It is possible to examine the technical and scale efficiency of each firm, relative to the frontier. With the addition of price information, allocative efficiency among firms can also be estimated and compared with each other.

While the DEA method constructs the efficiency frontier with the best performing farms of the sample, the method has to rely on specifying a production function and estimating its parameters using regression techniques. However, by assuming that all deviations from the frontier are the result of technical inefficiency, this simple deterministic model takes no account of the possible noise upon the frontier. The stochastic frontier model was then developed to account for noise and for agriculture's random nature. It assumes a double random error by adding to the deterministic model an additional random error.

Total factor productivity

In many studies, TFP growth is adopted as an indicator of agricultural competitiveness over time. TFP offers a number of advantages over other indicators: It is clearly definable, it is measurable using standard methods and it can be compared, to some extent, over time and across space and across various scales of study.[1] In addition, the ability to quantify economic benefits over time makes TFP a useful tool that can be incorporated into a cost/benefit analysis framework for policy analysis. These advantages make it the tool of choice to examine the effectiveness of policies designed to increase economic well being. However, estimating the relationship between policy measures (including R&D investments) and productivity offers some challenges, as there are often long lags between policy implementation and impacts on productivity and because there are other factors influencing productivity.

As Alston *et al.* (2010) and others have pointed out, TFP, which includes all inputs and outputs for a sector, is only a theoretical construction. Thus these authors prefer to use the term Multi-Factor Productivity (MFP) when referring to empirical studies to explicitly acknowledge that some outputs and inputs are inevitably excluded from the analysis due to data limitations. The two terms are used interchangeably in the literature.

TFP growth has three components which can be estimated using farm-level data (Coelli *et al.*, 2005).

- **Technological change** indicates a change in the technology available, used by the best performing farms that define the production frontier (technological progress from f to f' in Figure 2.2). Technological progress results from the adoption of innovations. There can also be technological regress.

- **Technical efficiency change** represents the movement of individual farms towards the efficiency frontier (technical efficiency increase from A to f in Figure 2.2). Technical efficiency (sometimes referred to as pure technical efficiency) assumes constant returns to scale and shows whether a firm is able to attain the maximum output from a given set of inputs.

- **Scale efficiency change** can be identified by the scale elasticity, calculated as the ratio between output and input growth rates for a given technology and is represented by a movement along the efficiency frontier due to a change in firm size (economies of scale from B to C in Figure 2.2 as C as a scale elasticity of 1, while B has a scale elasticity lower than 1). Scale efficiency gives insights into whether the firm operates at an optimal or sub-optimal size. Farms that are scale efficient operate under constant returns to scale (CRS) and have a scale elasticity of one, while scale inefficient firms could exploit scale economies or diseconomies.

Technological progress reflects advances in technology and adoption by early innovators, while technical efficiency increase reflects adoption of technology by individual farms, allowing them to move towards the frontier. The firms that previously defined the frontier are not necessarily those that adopt new technology available.

Figure 2.2. Pathway of productivity growth

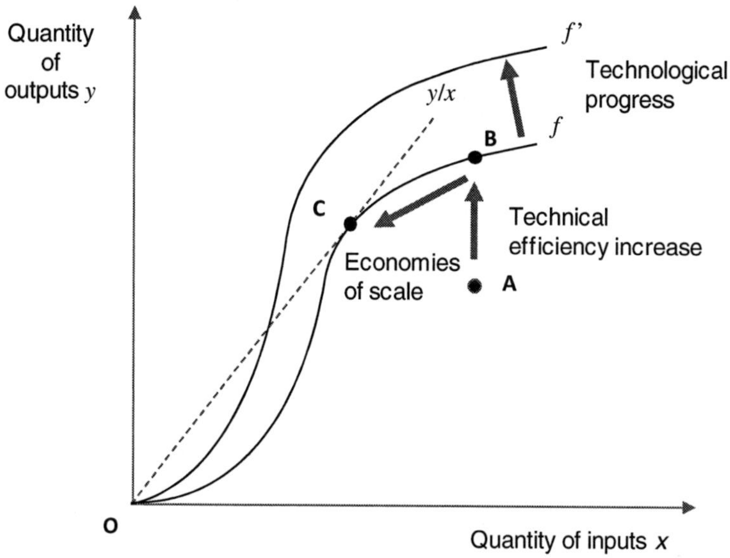

Source: Latruffe (2010) after Coelli *et al.* (2005).

Productivity and efficiency

Efficiency gives an indication of whether firms are able to use the existing technology in the best way.[2] It has three components: scale efficiency, technical efficiency and allocative efficiency. The first two, mentioned above, are components of productivity and refer to physical notions, theoretically independent of input and output prices.

The third component of efficiency, the allocative efficiency of a firm (also called its price efficiency), reflects its ability to use inputs in their optimal proportions given their respective prices, or to produce an optimal combination of outputs given their respective prices. A firm is allocatively efficient if its outputs and inputs maximise its profit (or minimise its costs) at given prices. Allocative efficiency implies technical efficiency, as in order to maximise its profits, the firm must firstly lie on the production frontier. However, technical efficiency does not necessarily imply allocative efficiency, since the combination of outputs and inputs can be optimal with respect to the production possibilities, but not be profit maximising.

Innovation

Innovation involves the introduction of new or significantly improved goods and services, or the use of new inputs, processes, organisational and marketing methods. These innovations can be new to the firm, new to the sector, or new to the world (OECD and EUROSTAT Oslo Manual, 2005). Firms innovate to improve productivity and profitability.

While specific forms of innovation can be easy to quantify in terms of adoption (e.g. the number of firms using solar-powered water pumps), it is far more difficult to develop general measures of innovation. For any given firm, industry or country, there are many possible forms of innovation that can interact in a myriad of ways. As such, aggregating different forms of innovation into a single measure is difficult. Given that increasing productivity is often the outcome of innovation, TFP growth is often used as a quantifiable measure of innovation in a firm, sector or country. However, innovation can help pursue objectives other than productivity growth, in particular in terms of product quality, diversity and safety, sustainability and animal welfare, which are generally not measured in volumes of agricultural outputs or inputs. This report considers specifically in Chapter 5 the impact of R&D on productivity growth, in addition to providing a brief overview of the nature of R&D.

Notes

1. However, it is difficult to compare TFP estimates from different studies, which are likely to rely on different data types and methods.

2. This notion of efficiency refers to the neoclassical efficient allocation of resources and the Pareto optimality criterion. Considering a firm that uses several inputs and produces several outputs, it is efficient in the way it allocates its resources if a reduction in any input requires an increase in at least one other input or a reduction in at least one output (Lovell, 1993).

Chapter 3

The evidence on agricultural productivity growth and competitiveness

In the last decade, agricultural productivity growth has decreased in many high level countries, but it is strong in Brazil, China and South Africa, as well as in major transition economies. Situations are contrasted in developing countries and overall, there is no widespread evidence that Total Factor Productivity (TFP) growth is slowing. Rates of TFP growth at European Union level are variable by member state, as is the contribution of technical efficiency and technological progress. In OECD countries, labour productivity increased faster than land productivity as farm labour declined faster than farm land. At the global level, the growth rate of crop yields has declined in the last 15 years compared to previous periods, but at a different pace across commodity. Evidence on competitiveness in the agricultural and agri-food sector based on trade or cost-related measures is relatively scarce.

The recent surge of interest for productivity growth is primarily linked with concerns about the ability of the sector to meet higher food demand from a growing and richer population in the longer term, as well as higher demand for non-food use, rather than competitiveness per se. Given that land, water and other inputs are not infinite, there is a consensus that productivity growth is necessary. In this context, there is growing interest in returns from R&D expenditures on productivity growth and more general debate on the role of government in innovation systems. At the same time, improving the productivity, efficiency and competitiveness of individual farms and the sector remains an important objective of agricultural policies in many countries.

Earlier work showed that productivity growth had strong links with competitiveness. Studies were often carried out in the context of greater exposure to regional or global competition following policy reforms, multilateral trade negotiations, regional free trade agreements or EU enlargement. Some studies looked at long-term trends of productivity growth, while others at developments in more recent periods. The purpose of many studies was to draw information on relative competitiveness through comparing productivity growth rates over the same period across countries. Generally speaking, macro-level data has been widely used to examine the long-term trends of productivity growth across countries, while more recently farm-level data has been used to compare across farm types.

This chapter focuses on the most recent studies, while the tables in Annex A summarise the coverage, choice of indicator, and main results of many studies reported in Latruffe (2010). Long-term developments in total factor productivity (TFP) are first reported, then results of studies decomposing TFP change into technological change and technical efficiency change, and evidence from partial productivity indicators. Finally, studies using measures of relative competitiveness across countries or agro-food sub-sectors other than productivity are briefly reviewed. Most studies focus on agriculture with only a few on the agro-food sector (i.e. food processing industries).

As there is no consensus on how to measure agricultural TFP across countries, estimation methods in the reviewed literature, as well as sources of data, vary widely. There is thus limited scope for cross-country comparisons of TFP estimations across sources mentioned, with the exception of a few studies where consistent data collection and methodology has enabled comparison over time and across countries (e.g. Ball *et al.*, 2010; Butault and Réquillart, 2010). In addition, there are also some accounting problems related to data collection for the TFP measurement, in particular capital and labour used on farm, and output. For example, TFP measures in many previous studies do not take account of those outputs related to agricultural production not valued in the market because these outcomes are externalities or have public good characteristics. They use output estimates that are reported in official statistics (e.g. national accounts). Similarly, changes in the quality of soil and water may affect productivity but are difficult to measure.[1]

Main developments in agricultural productivity reported in the literature

This sub-section focuses on most recent studies reporting long-term developments in agricultural productivity or comparisons across countries. It includes, in particular, a number of studies comparing the relative levels and growth rates of agricultural productivity across member states of the European Union and in the United States, using comparable aggregate data (e.g. Ball *et al.*, 1997, 2010; Butault and Réquillart, 2010); various recent ABARE publications (e.g. Sheng *et al.*, 2010, 2011); and evidence at global level collated from different sources and reported in Alston *et al.* (2010).

In summary, agricultural productivity levels are generally found higher in the United States than in most EU member states over the post-war period. Since 2000, productivity growth declines in both cases.[2] In Australia, the productivity growth of agriculture was strong between the mid-1950s and 2000, but has been fluctuating at a lower level since, partly due to a series of droughts that have affected the sector in 1994, 2003 and 2006. Agricultural productivity growth has resumed in transition economies following the post-1990s declines and is quite high in some countries. The situation in emerging and developing economies is diverse, with productivity growth particularly strong in Brazil and China in recent years.

A recent book by Alston, Babcock and Pardey (Alston *et al.*, 2010) provides a broad picture of global agricultural productivity as it looks at a broader range of countries and regions, using various indicators of total and partial productivity, data sets and estimation methods. In Chapter 4, Keith Fuglie presents agricultural TFP growth at global level estimated with the FAO data (Table 3.1), and in a large number of countries and regions (Table 3.2).[3] He does not find widespread evidence that TFP growth is slowing, although in many developed countries annual TFP growth is slightly lower in the 2000s than in the 1990s. While TFP growth may have been slowing in developed countries, rapid productivity growth has been recorded in several emerging economies, like Brazil, China and South Africa, and productivity is recovering fast in many transition economies, in particular of the former Soviet Union. Looking at TFP growth in agriculture of different countries and regions of Sub-Saharan Africa, Block (2010) finds that significant increases have been achieved since the 1980s, following declines recorded over the 1960s and 1970s. Yet, national situations are contrasted and land and labour productivities are still low by global standards.

Table 3.1. Productivity indicators for world agriculture, 1961-2007

Average annual growth rate by period (%)

Period	Output	Input	TFP	Output per worker	Output per hectare	Cereal yield (t/ha)
1961-69	2.81	2.31	0.49	0.96	2.39	2.84
1970-79	2.23	1.60	0.63	1.46	2.21	2.62
1980-89	2.13	1.21	0.92	0.97	1.72	1.00
1990-99	2.01	0.47	1.54	1.15	1.74	1.61
2000-07	2.08	0.74	1.34	1.72	2.10	1.01
1970-1989	2.18	1.40	0.77	1.22	1.97	2.31
1990-2007	2.04	0.59	1.45	1.40	1.90	1.35
1961-2007	2.23	1.24	0.99	1.25	2.01	2.02

Source: Table 4.6 in Chapter 4 of Alston *et al.* (2010), based on FAOSTAT data.

Table 3.2. Total Factor Productivity growth of agricultural in world regions, 1961-2007

Average annual growth rate by period (%)

	1961-69	1970-79	1980-89	1990-99	2000-07	1961-2007
All developing countries	**0.18**	**0.54**	**1.66**	**2.30**	**1.98**	**1.35**
Sub-Saharan Africa	0.36	-0.07	0.57	1.17	1.08	0.62
Latin America and Caribbean	0.29	0.70	1.20	2.54	2.60	1.47
- North East (mainly Brazil)	-0.52	-0.76	3.08	3.81	3.63	1.87
- Andean countries	1.45	0.59	1.01	2.73	1.74	1.49
- Southern Cone	0.36	1.73	0.03	2.15	2.03	1.27
Asia (except west)	-0.02	0.63	1.95	2.60	2.37	1.53
- North East (mainly China)	-0.12	0.30	2.77	4.08	2.83	2.03
- South East	0.68	2.26	0.98	1.78	2.59	1.66
- South	0.77	0.64	1.98	1.60	1.70	1.23
West Asia	1.06	0.00	2.82	2.25	2.04	1.64
North Africa	-0.10	0.61	1.33	1.46	0.95	0.89
Oceania	-0.20	0.07	-0.11	0.63	0.43	0.17
All developed countries	**1.21**	**1.52**	**1.47**	**2.13**	**0.86**	**1.48**
United States and Canada	0.86	1.37	1.35	2.26	0.33	1.29
North-West Europe	1.17	1.31	1.22	1.63	0.59	1.21
South-East Europe	1.56	1.46	1.91	2.03	0.82	1.59
Australia-New Zealand	0.93	1.29	1.26	0.53	-0.53	0.74
Asia (e.g. Japan, Korea)	-7.47	-0.86	0.39	1.59	1.80	-0.74
South Africa	0.50	1.53	1.80	2.75	3.09	1.95
Transition economies	**0.67**	**-0.26**	**0.25**	**0.73**	**1.92**	**0.61**
Central and Eastern Europe	0.63	0.38	0.60	1.92	-0.12	0.72
Former Soviet Union	0.73	-0.58	0.20	0.18	3.28	0.65
- Baltic	1.96	-0.79	0.51	0.23	2.28	0.61
- Central Asia and Caucasus	-0.56	1.85	-1.72	3.51	2.47	1.28
- Eastern Europe	1.23	-0.64	0.22	1.19	3.82	1.03

Source: Table 4.7 in Chapter 4 of Alston *et al.* (2010), using various authors' estimates.

Looking at relative productivity growth in the United States and eleven EU member states (Belgium, Denmark, Germany, Greece, Spain, France, Ireland, Italy, the Netherlands, Sweden and the United Kingdom) on a comparable basis, Ball *et al.* (2010) find that US agriculture had a higher TFP than European agriculture throughout the period 1973 to 2002, only with the exception of some countries over specific time periods (Belgium until 1985 and the Netherlands until 1992).[4] Only Sweden and Spain achieved faster rates of agricultural productivity growth than the United States, partly because they started from a lower base than other countries covered by the study. At the end of the period, there are still large differences in TFP among EU member states (from 59% of the 1996 US level in Ireland, to 94% in the Netherlands, compared to 105% in the United States). For an earlier period (1948-94), Ball *et al.* (1997) had investigated the evolution of agricultural productivity in the United States using the Fisher index method. They found that the estimated TFP for the United States increased at an average annual rate of 1.94% over the period and that the highest growth rates were found during 1966-69 (2.75%) and 1989-94 (2.87%), compared to less than 1% during before 1957. In an early period (1973-93) and for Denmark, France, Germany, the United Kingdom and the United States, Gopinath *et al.* (1997) find that TFP growth is the main driver of growth in agricultural value-added. The four EU member states had a higher growth in agricultural value-added than the United States, but their growth rate, which was very high in the 1970s, decreased in the later period. For a longer period, Alston *et al.* (2008a) report an average TFP growth of 1.56% per year during 1911-2002 for US agriculture. They also find that the highest TFP growth is for the period 1959-89 (2.11% per year compared to 1.24% during 1911-49), but that it is lower during 1990-2002 (1.01% per year).

Using similar data and methodology but for a longer time period (1959-2008 instead of 1973-2002),[5] Butault and Réquillart (2010) relate developments of agricultural production in France and the United States to changes in the partial factor productivity. French agricultural output grew strongly (+2.1% per year) between 1960 and 1979 and became more intensive, with an annual TFP growth rate of 1.5%. From 1979 to 1996, output growth slowed (1.1% per year) but the partial productivity of intermediate inputs and capital, and the TFP growth still increased (1.9% per year). Since the mid-1990s, lower TFP growth has been observed (0.6% per year), which can be used to explain the stagnation of French agricultural production in recent years. While labour productivity continued to increase, intermediate input productivity remained stable and capital productivity decreased.

In France as in many other members of the European Union, output growth and intermediate input productivity growth were lower in the 2000s than in the 1990s. It was even negative in some countries like Ireland. This study finds that over the whole period 1959-2008, the average annual growth rate of agricultural productivity in France was lower than in the United States (1.4% compared to 1.8%). As in the United States, the reduction in agricultural production in France was accompanied by improvements in productivity, notably of intermediate inputs.

Butault and Réquillart (2010) discuss potential causes for the decline in agricultural production and productivity growth in French agriculture. While the slowing of production growth can be attributed to the decrease in output prices resulting from successive reforms of the Common Agricultural Policy (CAP), the decline in productivity growth cannot be explained by those reforms. In contrast, the strengthened competition due to the CAP reforms should have increased efficiency in the use of inputs. The stagnation of wheat yields in France since the mid-1990s and the deterioration of intermediate input productivity may indicate that technical progress has halted or that input-saving technologies are not adopted because they imply too large a change in production techniques.

Butault and Réquillart (2010) also examine productivity changes in agro-food (processing) industries in France and seven other EU member states. The TFP of French agro-food industries decreased over 1997-2002 and remained stable between 2002 and 2006. The authors suggest this poor performance could be explained by the low productivity of intermediate inputs (mainly agricultural products), which account for over three-quarters of gross output value, difficulties in improving the conversion of intermediary inputs into output, and maybe more stringent health and environmental regulations. They note that French agro-food industries include many very small enterprises and few large ones. They also find that TFP in agro-food industries decreased in six EU member states examined over 1992-2006, including France, but increased in the Netherlands and the United Kingdom.

Latruffe (2010) report a number of studies investigating partial or total factor productivity, and profitability in the food manufacturing sector but most studies in this review (summarised in Annex Table A.1) cover periods prior to 2000. Again, the purpose of those studies is to compare the performance of agricultural production across countries and product sectors, and/or to monitoring agriculture developments in a specific country. Overall, TFP increases in the food processing industry, faster in the United States than in other countries (Ruan and Gopinath, 2008; Gopinath, 2003; Buccola *et al.*, 2000; Chan-Kang *et al.*, 1999).

A series of recent studies report trends in productivity growth in Australian broadacre agriculture (e.g. Sheng *et al.*, 2010 and 2011; Mullen, 2010).[6] Broadacre productivity growth almost tripled between 1953 and 2000. From an index of 100 in 1953, broadacre productivity grew to 218 in 2007, peaking at 288 in 2000. TPF has been variable over the period, but more so at the end with significant drops in 2003 and 2006 linked to adverse climatic conditions. Sheng *et al.* (2010) identify a structural break in productivity in the mid-1990s: broadacre productivity grew at about 2.2% per year before 1994, but that rate declined to 0.4% a year thereafter. According to the analysis, climatic conditions alone do not explain the totality of the slowdown, and the slower growth in public R&D expenditures from the 1970s also played a role.[7] Between 1978 and 2007, TPF growth varied across the broadacre industries and between states (Nossal and Sheng, 2010, Table 1). Crop specialists had consistently achieved higher rates of TFP growth but were the most affected by recent declines. In contrast, the beef industry was more resilient.

Latruffe (2010) provides an overview of studies carried out in the 1990s and 2000s that compare the performance of different countries or different farm types using total factor productivity (Annex Table A.2). These studies include earlier work by Ball *et al.* (1997, 2001 and 2006) and many studies on agricultural productivity developments in European countries. The studies of agricultural productivities in European countries cover shorter periods because they use farm-level data. Many of these studies decompose productivity change into technical efficiency change and technological change and are described in the following chapter, and summarised in Annex Table A.3. Some also make comparisons across farm types.

Decomposition of productivity changes

Studies comparing productivity and productivity changes across EU member states note that over time, there has been convergence with productivity growing faster in countries that initially had a low level. However, some countries still lag behind. Along this line, some studies have paid attention to the new entrants to the European Union and found that some of these countries had high technical efficiency (e.g. dairy farms in Poland), though their TFP level is generally lower than many EU15 member states (see, for example, Brummer *et al.*, 2002; Fogarasi and Latruffe, 2009).

Using Data Envelopment Analysis (DEA) to compute Malmquist TFP indices, Rungsuriyawiboon and Lissitsa (2006) find that TFP grew by 1.3% per annum in the EU15 over the period 1992-2002, mainly due to an increase in technical change (1.36%), while technical efficiency, which shows whether a firm is able to produce the maximum output from a given set of inputs, decreased by 0.11%. Over the same period, annual TFP growth is slightly higher in the EU10 (1.4%) due to increases in both technical efficiency (1.12%) and technological change (1.3%), while scale efficiency[8] slightly decrease (-0.002%). Again there is a high diversity among member states, with TFP growth ranging from -0.5% in Ireland to 5.3% in Estonia. This study also covers a wide range of transition economies. Applying the same method over a longer period (1980-2000), Coelli and Rao (2005) find that TFP growth in EU15 member states is mostly due to technical change, while in new member states technical efficiency change `has made a significant contribution to the TFP growth.

Carroll *et al.* (2009) calculate TFP growth for several production types in Ireland during 1996-2006 (2000-06 for sheep): the average annual growth rate over the period was 2% for cattle rearing, 1.4% for dairy, 0.9% for cattle finishing, 0.4% for sheep and -0.2% for cereals. They also report technical efficiency change, which show different pattern over time. All production types experienced the worst technical efficiency change (deterioration) in the first periods and the best change (progress) in the last periods. When technical efficiency change is averaged over all types of farms, figures indicate that the strongest deterioration was in the first period 1996/97 (-1.564%) and the strongest progress was in the last period 2005/06 (1.365%).

Looking at different farm types in England and Wales, Barnes *et al.* (2010) find that technical efficiency is relatively high for most farm types and was relatively stable during the period 1989-2008 despite market disturbances, in particular due to animal diseases and policy reforms. In a similar study covering the period 1982-2002, Hadley (2006) found there was no change in technical efficiency for cereal and poultry farms, or decreases for general cropping and mixed farms. He calculated technological change and found it was always positive, with cereal and mixed farms experiencing strongest progress (5.8% and 5.2% respectively) and poultry farms the smallest progress (1.6%).

Developments in partial productivity measures

This chapter presents estimates of partial factor productivity calculated by the OECD Secretariat and found in the literature. The OECD land and labour productivity estimated by the OECD in this study covers the period 1969 to 2008. The statistics show that real output increased in all OECD countries examined except Japan (Table 3.3). The data indicates that both farm labour and land have decreased in most of OECD countries, with farm labour declining at a higher rate than land in most countries. As a result, labour productivity growth rates tend to be higher than land productivity growth rates in most countries.

Figure 3.1 shows there are wide differences in land and labour productivity as well as land-labour ratio among OECD countries. The horizontal axis in Figure 3.1 represents labour productivity as the value of agricultural output (expressed as average purchasing power parity between 1999 and 2001) per economically active worker in agriculture.[9] The vertical axis represents land productivity as the value of agricultural output per 1 000 hectares of agricultural land. The horizontal and vertical axes are scaled as logarithm so that each diagonal (45 degree line) plots the same land-labour ratio. Land productivity is particularly high in those countries with scarce land endowment relative to labour (e.g. Korea and Japan). The initial land endowment in Korea and Japan was respectively 0.5 and 0.7 hectare per worker in 1969, whereas the agricultural worker in Australia and Canada was endowed with 1 155 and 114 hectare of land, respectively. Labour productivity tends to be higher in countries where land is abundant relative to labour.

Figure 3.1. Developments in land and labour productivity in OECD countries, 1969-2008

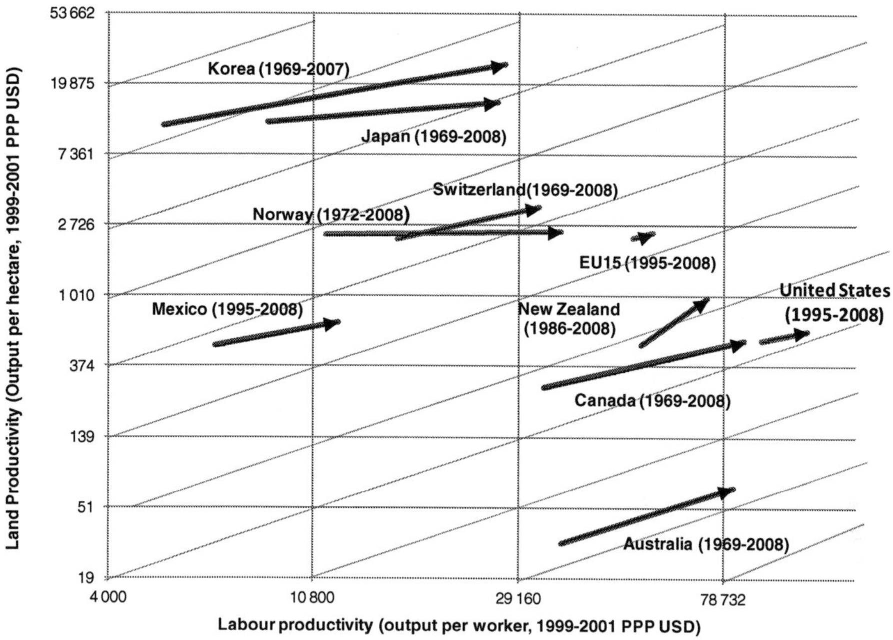

Notes:

Differences in productivity level and trend across countries reflect to a large extent differences in land endowment.

Labour is measured as the number of economically active worker. Land is the sum of area harvested and in permanent pastures

The start and end points of the arrow represent labour and land productivity during the average of initial and last three years (1969-71 and 2006-08).

Source: Secretariat's calculations using FAO Stat and ILO Laborsta data. US employment data from National Agricultural Statistical Service. Swiss and Luxembourg employment data from official estimates.

Table 3.3. Developments in output, land, labour, factor productivity and land per worker, 1969-2008

	Period	Output	Farm labour	Farm land	Productivity of		Ratio land to labour	
					Labour[1]	Land[2]	First year	Last year
		Annual growth rate (%)					Ha per worker	
Australia	1969-2008	2.5	-0.4	-0.3	3.3	3.2	1 155	1 176
Canada	1969-2008	2.5	-0.9	-0.1	5.1	2.6	114	168
Japan	1969-2008	-0.2	-1.8	-0.8	5.7	0.8	0.7	1.7
Korea	1969-2007	2.4	-1.6	-0.6	11.1	3.8	0.5	1.0
Mexico	1995-2008	2.8	-2.0	-0.3	6.5	3.3	14	18
New Zealand	1986-2008	1.7	-0.3	-1.4	2.1	4.5	101	75
Norway	1972-2008	0.7	-1.8	0.3	7.3	0.3	5	15
Switzerland	1969-2008	0.4	-1.1	-0.7	2.6	1.6	7	9
United States	1995-2008	1.4	-0.6	-0.2	2.2	1.6	178	190
EU15	1995-2008	0.3	-1.4	-0.6	2.0	0.9	22	24

1. Value of agricultural output per economically active worker at constant price.
2. Value of agricultural output per hectare at constant price.

Source: Secretariat calculations based on FAO Stat and ILO Laborsta data; US employment data from National Agricultural Statistical Service; and Swiss and Luxembourg employment data from official estimates.

Table 3.3 suggests that productivity growth in many OECD countries has been driven by a significant improvement in labour productivity, leading to higher land endowment per worker. Moreover, initial resource endowment characterises not only the level of land and labour productivity, but also its growth path. Figure 3.1 shows that labour productivity growth is generally higher than land productivity growth in countries where initial land endowments are relatively scarce. This trend is captured by horizontal growth paths (e.g. Korea, Japan, Norway and the European Union). The labour and land productivity in Norway grew respectively at 7.3% and 0.3% annually between 1972 and 2008, compared to 5.7% and 0.8% in Japan for the same period. In the European Union, the gap between labour and land productivity growth is much larger between 1995 and 2008, namely 2.0% and 0.9% respectively. Land endowment per worker generally increased over the reviewed period, but to various degrees (Table 3.3).

Land and labour productivity in Australia increased along a diagonal line, meaning that land productivity and labour productivity grew at similar rates of around 3.2% annually between 1969 and

2008, while land endowment per worker remained around 1 200 hectares (Table 3.3). In Canada, labour and land productivity grew respectively at 5.1% and 2.6% annually between 1969 and 2008, leading to an increase in the land to labour ratio from 114 to 168 hectares. In New Zealand, the annual land productivity growth rate was 4.5% annually between 1986 and 2008, which exceeded the labour productivity growth rate of 2.1% annually. Land endowment per worker declined from 101 to 75 hectares during the same period. This means that land is worked more intensively.

Alston, Beddow and Pardey in Chapter 3 of Alston *et al.* (2010) contain a similar analysis of developments in partial factor productivity for regional and economic groups of countries worldwide for the period 1961-2005 (Table 3.4). It shows, for example, that in the Latin America and Caribbean region, the labour productivity increased faster than the land productivity, while Sub-Saharan Africa has become more labour intensive so its land to labour ratios have declined. Despite significant increases in land productivity in low income countries and Sub-Saharan Africa, it remains lower than in most other country groupings, except Australia and New Zealand which have low land productivity but very high labour productivity.

Alston, Beddow and Pardey (2010) also indicate that the growth in the land and labour productivity of agriculture decelerated after 1990 in the United States. For example, the growth rate of labour productivity was 2.38% and 4.11% each year during 1911-49 and during 1949-89 respectively, while it was only 1.59% each year during 1990-2006. The slowing down of land and labour productivity after 1990 is widespread across countries (Table 3.4).

Block (2010) compares developments in land productivity and labour productivity in different regions of Sub-Saharan Africa (SSA) between 1961/65 and 2006/07. Compared to other regions and the world average, partial factor productivity growth in Sub-Saharan Africa is low and has been mainly driven by increased yield per hectare with little growth of output per worker, reflecting the impact of growth in population. There are, however, large differences in levels and rates by SSA region. From a low level in 1961/65, Southern Africa achieved very high increases until 1981/85 when partial factor productivity decreased and then resumed slower growth. West Africa made substantial progress in increasing labour productivity from the 1980s. Consistent progress was also made in

East Africa but at a slower rate, while Middle Africa experienced slow declines in labour productivity.

Table 3.4. Global growth in agricultural land and labour productivity, 1961-2005

Average annual growth rate by period (%)

	Land productivity		Labour productivity	
	1961-90	1990-2005	1961-90	1990-2005
World	2.03	1.82	1.12	1.36
- excluding China	1.90	1.19	1.21	0.42
Latin America	2.17	2.83	2.15	3.53
Asia	2.56	3.01	1.83	2.72
- excluding China	2.45	1.83	1.69	1.24
- China	2.81	4.50	2.29	4.45
Africa	2.18	2.21	0.68	0.90
Per capita income				
High	2.00	2.39	0.46	1.03
Middle	2.35	2.30	1.51	2.02
Low	1.61	0.72	4.26	4.18
Top 20 producers	2.11	2.16	1.17	1.77
- excluding China	1.98	1.38	1.33	0.63

Source: Table 3.6 in Chapter 3 of Alston *et al.* (2010).

Looking at developments in crop yields, Alston *et al.* (2010) also find evidence that land productivity growth has been uneven across agricultural commodities. The slowing growth in cereal grain yields identified by the *World Bank Development Report 2008* (World Bank, 2007) is illustrated in Table 3.5. It shows the slowing down of annual average yield growth rates for maize, wheat, rice and soybeans globally and for most country groupings, except in Eastern Europe for wheat and soybeans. Alston *et al.* (2008) also report similar developments in developing and developed countries. For example, maize yield growth was 2.53% per year during 1961-89 and 1.92% per year during 1990-2006 for developing countries, while 2.50% and 1.67% per year respectively in developed countries. Similar cross-country disparities in the growth rates of yield per hectare between the two time periods are also shown for wheat and rice. They also note that slowing cereal yields are found in the majority of large producing countries.

Table 3.5. Global yield growth rates for selected crops, 1961-2007

Average annual growth rate by period (%)

	Maize		Wheat		Rice		Soybeans	
	1961-90	1990-2007	1961-90	1990-2007	1961-90	1990-2007	1961-90	1990-2007
World	2.2	1.77	2.95	0.52	2.19	0.96	1.79	1.08
North America	2.2	1.4	2.23	0.01	1.67	1.54	1.05	0.04
Western Europe	3.3	1.81	3.31	0.63	0.38	0.55	1.64	0.05
Eastern Europe	1.91	0.97	3.18	-1.69	-0.41	1.07	1.9	2.29
Per capita income								
High	2.34	1.48	2.47	0.06	1.07	0.54	1.14	0.02
Middle	2.41	2.12	3.23	0.85	2.54	0.81	3.21	2.08
Low	1.07	0.65	1.32	2.15	1.46	2.16	2.63	0

Looking at developments in EU cereal yields, OECD (2011) finds that they have increased in the EU15, at a rate of 1.5% per year between 1990 and 2009 (from 4.7 tonnes per hectare to close to 6 tonnes per hectare). In new member states they are lower (less than 4 tonnes per hectare) and more variable mainly due to extreme climatic events, but they recovered their pre-1990s level in 2008. Milk yields have also increased in most member states in the 1990s and 2000s, and there is no generalised slowing-down pattern of growth rates in the most productive EU15 countries.

The same report (OECD, 2011) also notes that the productivity of intermediate inputs in the European Union has increased continuously during the period 1995-2005, but has then stagnated until 2008, as productivity growth in some member states has been compensated by declines in others.[10]

Latruffe (2010) also reports a number of studies using partial factor productivity to compare across countries, sectors or time (Annex Table A.4). For example, Alston *et al.* (2008a) finds that maize and wheat yield growth slows down after 1990 and that maize yield growth is lower in developing countries than in developed countries after 1990. Mulder *et al.* (2004) finds that land and labour productivity of selected protected commodities was lower in Mercosur countries than in the European Union in 1995, but that costs of productions were lower. When comparing cereal yields and labour productivity in six EU15 member states over 1996-2000,

Thorne (2005) found that Italy was lagging behind. Looking at cow milk yields in 2006 in EU member states and candidate countries, Van Berkum (2009) found that only the Czech Republic, Hungary and Estonia were close to the EU15 average and that the Balkans had the lowest yields.

Evidence on competitiveness in agriculture and the agro-food sector based on measures other than productivity

Productivity is often used as an indicator of competitiveness but other indicators are also used. Latruffe (2010) reports a number of studies using trade-related measures such as Revealed Comparative Advantage (RCA) index to measure competitiveness,[11] while others use strategic management index such as production costs or profitability indicators to measure competitiveness.

Trade measures of competitiveness generally focus on agri-food sectors. Again, many studies make comparisons between EU member states and with a large range of trade partners such as Australia, Canada, the United States, Mercosur countries (notably Brazil), Russia, Ukraine and other European countries that are not members of the European Union. One purpose is to identify sectors in which a country has or is gaining competitive advantage, notably in the context of EU accession. As results can be sensitive to the indicator chosen, many studies base their assessment on several indicators. Some use a large range of trade indicators and cluster analysis to rank countries (Annex Table A.5).

As shown in Annex Table A.6, Domestic Resource Cost (i.e. the ratio of opportunity costs of domestic production over the value added it generates) is widely used to assess the competitiveness of EU new member states, Russia and Ukraine. All studies but one concern the 1990s. They find that crop production is generally more competitive than livestock production, probably because of lower capital requirements in a period where capital was difficult to obtain. For crop production, some of those countries were more competitive than in the European Union and even at world level.

Studies on costs of production reported in Latruffe (2010) are generally outdated.[12] For the commodity examined in the studies reported in Annex Table A.7, production costs were generally higher in the European Union than in Brazil, other Mercosur countries and the United States. In the mid-1980s, costs of production were higher in the United States than in Canada.

Latruffe (2010) reports two studies which refer to competitiveness using profitability measures (Annex Table A.8). These are in fact indicators of farm-level performance which were not the focus of her review, but are regularly monitored in many countries.

Notes

1. Specific examples are mentioned at the beginning of Chapter 5.
2. Evidence on this relative decline depends on the data, the estimation method and the precise time period.
3. Where possible, he compares his results with other national studies and finds that they track relatively well TFP growth estimated by Ball *et al.* (2010) for 11 members of the European Union and the United States, and by Fan *et al.* (1999) for India, but his estimations are lower than others for Brazil in the most recent period, for Indonesia, and in particular for China.
4. This study uses data from agricultural accounts published by the Eurostat and the US Department of agriculture. Differences in productivity growth between EU member states and the United States are expressed using price indices. This is equivalent to the more familiar quantity index if revenue is assumed to equal costs in each period.
5. This study uses the same TFP measure but uses different weights to obtain an aggregate index of land, labour and capital.
6. Measures of productivity are based on data from ABARE farm surveys of broadacre agriculture, including grazing and cropping industries, and since 1989, the dairy industry. They use a gross output of production approach.
7. According to the analysis in Sheng *et al.* (2010), public R&D expenditures for agriculture have increased but at a lower rate than productivity. That is, there has been a long-term slowdown in growth in public R&D expenditure since the 1970s.
8. Scale efficiency gives insights into whether the firm operates at an optimal or sub-optimal size (Latruffe, 2010).
9. The economically active population comprises all persons of either sex who furnish the supply of labour for the production of goods and services during a specified time-reference period. It including both self employed and hired labour. For more information, see the ILO website (*laborsta.ilo.org/*).
10. See Figure 2.12 and Annex Table B.4 in Latruffe (2010).

11. Ratio of the share of trade (T) in commodity i over all commodities in a country (j) and the same share in all other countries (n), i.e. (Ti,j/Tj) / (Ti,n/Tn).

12. Some countries like the United States or Ireland, publish regular estimates of costs of production from survey data, while costs of production by commodity are being estimated in a number of EU member states as part of a EU research project, FACEPA (see FACEPA website at: *www2.ekon.slu.se/facepa*).

Chapter 4

Developments in public and private research and development efforts

Expenditure on research and development (R&D) is often used as an indicator of efforts in this area. This chapter describes trends in public and private expenditures on agricultural R&D, using OECD and ASTI databases. Most expenditures on agricultural R&D are made by the public sector, but the share of private expenditures is increasing in some OECD countries. Public R&D expenditure as a percentage of agricultural GDP ranges from less than 0.3% to over 4% among OECD countries. It is also very diverse in developing countries.

Organised R&D in the public and private sectors is the main source of new agricultural technologies that is expected to foster agricultural productivity growth over the long term. The traditional linear model of innovation postulates that innovation starts with basic scientific search, followed by applied research, and ends with product and process diffusion (Bush, 1945). The primary motivation of public investment is to fund and perform basic research that often has a long time horizon and carries high risks with uncertain returns (OECD, 2010). R&D involves effort to increase the stock of knowledge. Ideas from the larger stock of knowledge are incorporated through commercialisation and adoption into production technologies, good and services, and other forms of innovation.

Enhancing the flow of knowledge and technologies to end-users is critical for the adoption of technologies and knowledge that will lead to productivity growth. In agriculture, governments usually fund some extension activities to facilitate the adoption of new technologies.[1] Many OECD countries provide public extension services at the national or local level to diffuse new technologies and knowledge created by public research institutions. More recently the role of the private sector has increased either in the form of public and private service or fully commercial service (Swanson and Rajalahti, 2010). In some countries, farmers' organisations such as agricultural cooperatives are providing an extension service to the member farms. Input suppliers also play an important role in promoting the adoption of new technology, as well as processors, in particular in the context of vertical coordination arrangements.

Trends in public expenditure on agricultural R&D

Although private expenditures on agricultural R&D have become increasingly important in OECD countries, governments continue to be the main source of funds for agricultural R&D. In many developing countries, the public sector funds almost all agricultural R&D. Public expenditure on agricultural R&D includes financial support to R&D activities by both public and private institutions. Many OECD countries fully finance agricultural R&D undertaken in public institutions although in some countries (e.g. Australia and the Netherlands) this is co-financed by the public and the private sectors. Public funding typically continues to support basic research and some aspects of applied research where the incentives for private research are limited. In addition, the public sector often supports private research through the use of tax incentives, levies, matching grants, etc. As a result of the range of institutions involved, measuring R&D investment is often a formidable task.

OECD has been collecting information on R&D efforts by member countries (e.g. R&D expenditures and personnel) on a regular basis as a part of its main science and technology indicators.[2, 3] The standard measure of R&D expenditure is defined as the Gross Domestic Expenditure on Research and Experimental Development (GERD), which is further categorised by the four sectors of performance (Government, Business Enterprise, Higher Education and Private Non-Profit). GERD is further categorised by its socio-economic objective according to NABS 2007 classification.[4] Agriculture is one of the socio-economic objectives that includes R&D expenditures on agriculture, forestry, fisheries and foodstuff production. Data covers all OECD countries and some non-member countries (Argentina, Romania, Russia and South Africa). While most countries provide information on overall R&D expenditure at national level, agricultural R&D is less well covered. In some countries it is only available for some years and in others like the United States, it is completely missing. The US information reported below is taken from a ERS/USDA's publication on public and private funding of agricultural research and development.[5] To collect R&D expenditure, the OECD provides guidelines for collecting information (the Oslo Manual) and a common questionnaire, but it is dependent on the availability and quality of data in member countries.

Another source of information on agricultural R&D is the Agricultural Science and Technology Indicators (ASTI) database, managed by the International Food Policy Research Institute (IFPRI). The ASTI database mainly covers agricultural R&D data in emerging and developing economies, including some Enhanced Engagement countries for the OECD (China, Indonesia and South Africa). The definition and categorisation of agricultural R&D is generally in line with the OECD's R&D database.[6] Table 4.1 presents the evolution of agricultural public R&D expenditure as an annual growth rate for three periods: 1980s, 1990s and 2000s. Many countries have increased real public expenditure on agricultural R&Ds since the 1980s, in particular, the United States, Spain and

Japan. Over the 2000s, Austria, Ireland, Mexico, Slovenia, Spain and Norway recorded particularly higher growth rates of public expenditure in agricultural R&D at more than 5% annually. In Australia and in Portugal, real expenditure on public agricultural R&D declined significantly after 2000, although in Australia, this has been partly offset by expenditure by the private sector (Table 4.2).[7]

Table 4.1. Evolution of expenditure on public agricultural R&D, 1981-2008

Country	1980s		1990s		2000s	
	Period	Annual growth rate (%)[1]	Period	Annual growth rate (%)[1]	Period	Annual growth rate (%)[1]
Australia	1981-90	-0.3	1990-00	0.6	2000-08	-8.5
Austria	1981-89	0.9	1989-98	-3.1	1998-2007	6.5
Ireland	1981-89	-5.2		n.a.	2002-08	6.1
Israel[2]		n.a.	1993-99	3.5	2000-09	-1.1
Italy	1981-89	9.4	1990-99	-1.7	2000-08	3.1
Japan	1981-89	0.9	1991-99	1.4	2000-08	0.5
Korea		n.a.	1995-99	-2.5	2000-08	1.1
Mexico		n.a.	1993-95	-1.8	2002-03	7.5
Norway	1981-89	8.1	1989-99	-0.4	1999-2003	5.9
Portugal	1982-89	7.7	1990-99	7.2	2000-08	-11.1
Slovenia		n.a.	1994-99	2.7	2000-08	7.5
Spain	1981-89	11.9	1990-99	1.3	2000-08	5.4
Sweden	1983-89	-1.3	1989-99	-7.5	1999-2007	-1.9
Switzerland	1981-92	1.8	1992-00	0.2	2000-08	-0.3
United States[3]	1981-89	0.5	1990-99	0.1	2000-08	0.2

n.a.: Not available.

1. Calculated from expenditures in constant 2000 USD.
2. The statistical data for Israel are supplied by and under the responsibility of the relevant Israeli authorities. The use of such data by the OECD is without prejudice to the status of the Golan Heights, East Jerusalem and Israeli settlements in the West Bank under the terms of international law.
3. National data. The increase in expenditures over the 2000s stops in 2006 and expenditure decrease in the following two years.

Source: Secretariat's calculations based on OECD R&D Data.

Table 4.2. Evolution of private expenditure on agricultural R&D, 1992 to 2008

Country	1980s		1990s		2000s	
	Period	Annual growth rate (%)[1]	Period	Annual growth rate (%)[1]	Period	Annual growth rate (%)[1]
Australia		n.a.	1991-99	6.4	2000-08	8.8
Iceland		n.a.	1990-99	7.9	2000-08	10.1
Korea		n.a.	1995-99	-0.7	2000-08	11.1
Norway	1981-91	13.2	1991-99	3.1		n.a.
Spain	1981-89	22.3	1990-99	4.8	2001-08	11.4
United States [2]	1981-89	2.3	1990-98	2.0	1999-2009	n.a.

n.a.: Not available.

1. Calculated from expenditures in constant 2000 USD.
2. National data.

Source: Secretariat's calculations based on OECD R&D Data.

Another way of comparing the intensity of agricultural R&D across countries over time is the ratio of R&D expenditure and agricultural GDP. Using the OECD database, Figure 4.1 presents the evolution of public agricultural R&D expenditure as a percentage share of agricultural GDP in 1992, 2000 and 2006. In these countries (with the exception of Australia), the intensity of public R&D for agriculture increased over time. In Australia, the decrease in the intensity of public R&D was partly offset by the increase in private R&D expenditure (Figures 4.1 and 4.5). A wide diversity exists in the intensity of public R&D expenditure among OECD countries (Figure 4.2). Public agricultural R&D expenditure in the United States, Ireland, Iceland and Japan accounted for more than 3% of agricultural GDP in 2006, whereas less than 1% of agricultural GDP was spent for agricultural R&D in Austria, the Slovak Republic, Slovenia and Turkey.

Figure 4.1. Developments in the intensity of public R&D expenditure on agriculture in selected OECD countries, 1992, 2000 and 2006

Public R&D expenditures as a percentage of agricultural GDP[1]

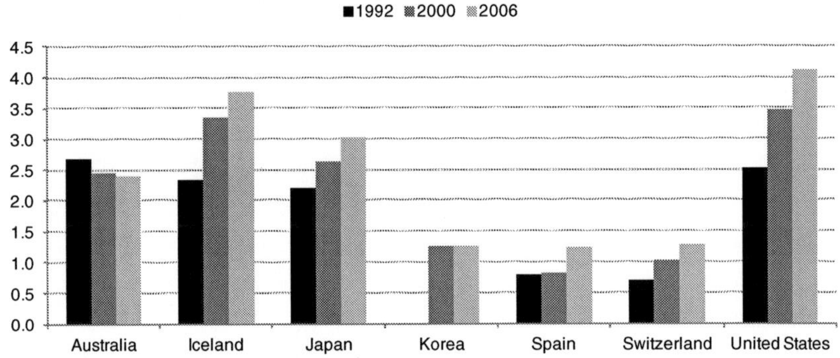

1. Agriculture includes crop, livestock, hunting, forestry and fishing.
Source: OECD R&D Database and US national database.

Figure 4.2. Intensity of public R&D expenditure on agriculture, 2006

Public R&D expenditures as a percentage of agricultural GDP[1]

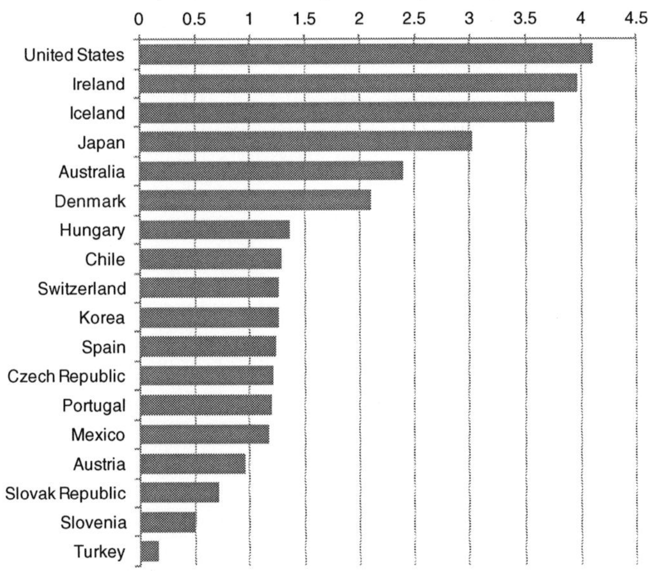

1. Agriculture includes crop, livestock, hunting, forestry and fishing.
2. Data from the OECD database except for Chile and Mexico, where they come from the ASTI database.
Source: OECD R&D Database and IFPRI/ASTI Database.

Figure 4.3. Evolution of public R&D expenditure on agriculture in selected non OECD countries, 1992, 2000 and 2005

Government expenditures as a percentage of agricultural GDP[1]

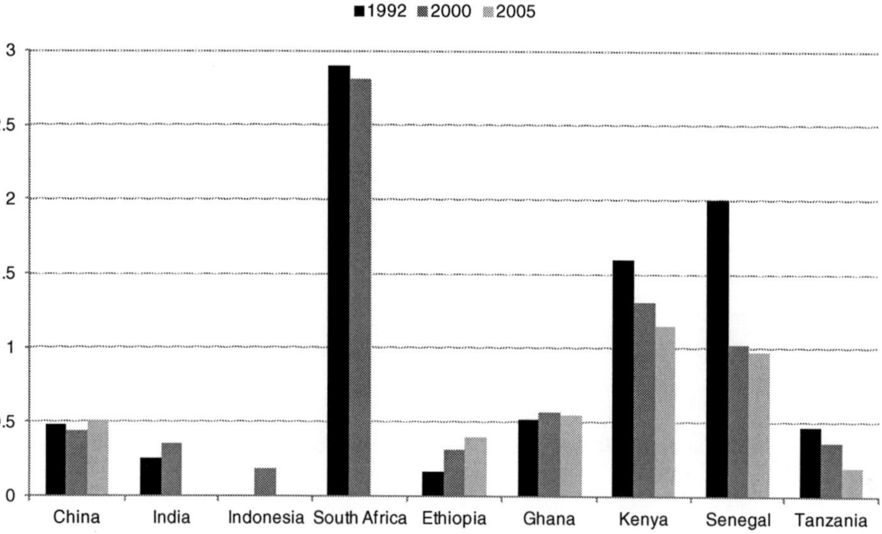

1. Agriculture includes crop, livestock, hunting, forestry and fishing.

Source: IFPRI/ASTI database.

In developing countries, the intensity of public R&D expenditure on agriculture is generally lower than in OECD countries. Among the nine countries where data are available in 2000, only South Africa, Kenya and Senegal spent more than 1% of agricultural GDP on public R&D. In China, Ethiopia, India and Tanzania, public expenditure on R&D accounted for less than 0.5% of agricultural GDP in 2005. In contrast to many OECD countries, the share of public R&D expenditure in agricultural GDP has not increased over time and has even fallen in several countries. For example, the intensity of public R&D in agriculture declined significantly in Kenya, Senegal and Tanzania between 1992 and 2005.

Figure 4.4. Share of public R&D expenditure in total agricultural support, 1986-88 and 2007-09

R&D expenditures as a percentage of Total Support Estimate

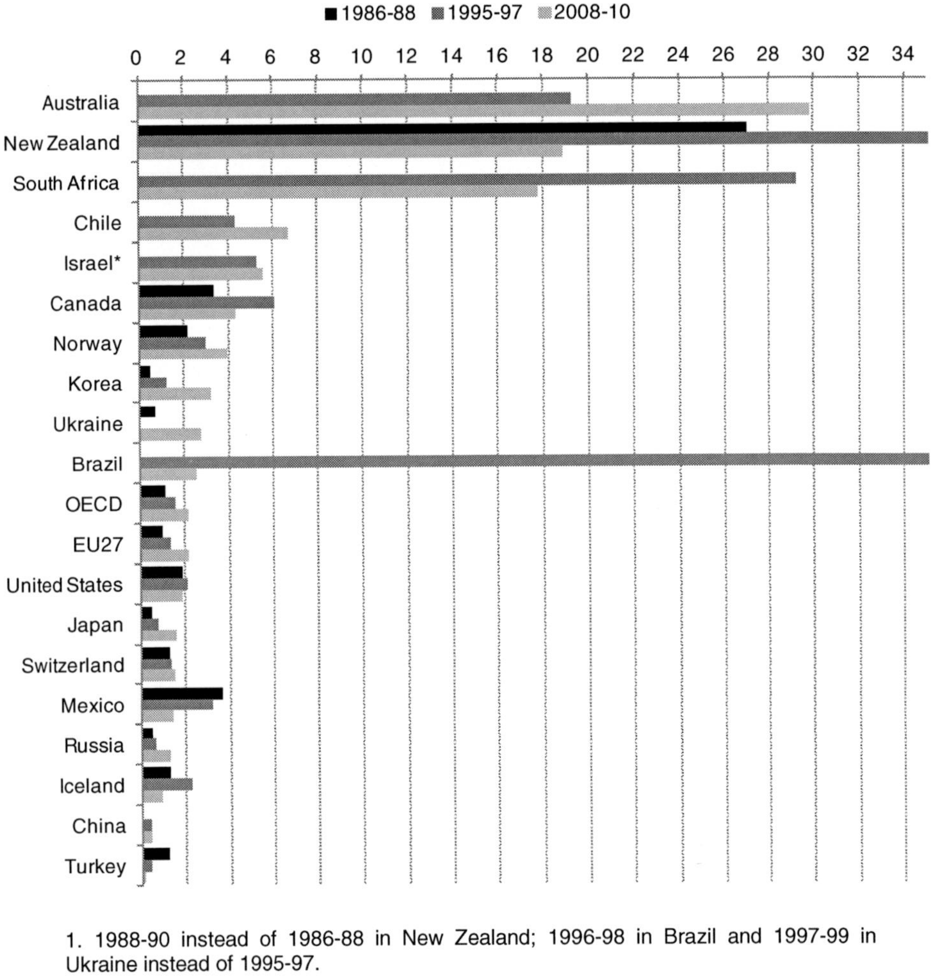

1. 1988-90 instead of 1986-88 in New Zealand; 1996-98 in Brazil and 1997-99 in Ukraine instead of 1995-97.

* The statistical data for Israel are supplied by and under the responsibility of the relevant Israeli authorities. The use of such data by the OECD is without prejudice to the status of the Golan Heights, East Jerusalem and Israeli settlements in the West Bank under the terms of international law.

Source: OECD PSE database, 2011.

To monitor and evaluate agricultural policies in member countries and selected emerging economies, the OECD estimates annual government budgetary expenditures on agricultural R&D activities. This information is contained in the Producer Support

Estimate (PSE) database,[8] as a part of the General Services Support Estimate (GSSE).[9] Government R&D expenditures on agriculture in the PSE database are lower than those in the OECD R&D database, which includes hunting, forestry and fisheries in the "agriculture" aggregate. In the OECD area as a whole, the relative importance of support to agricultural R&D in total agricultural support has increased over time. The share of budgetary expenditure on agricultural R&D in the total support estimate (TSE) increased from 1.2% to 2.2% between 1986-88 and 2008-10 (Figure 4.4). In fact, the share of R&D expenditure in agricultural support has increased in most OECD countries over the whole period, although in Canada, Mexico, New Zealand and the United States, it is lower in 2008-10 than in the mid-1990s. Australia, New Zealand, Israel, Canada, Norway and Korea have relatively higher shares of agricultural R&D expenditure in agricultural support than the OECD average in 2008-10. In Australia and New Zealand the share of agricultural R&D expenditure in agricultural support is particularly high as the level of total support to agriculture is much lower than the OECD average. While as a percentage of the total value of agricultural production, Australian agricultural R&D expenditure remains higher than the OECD average (1.4% in Australia compared to 0.8% as the OECD average in 2008-10), this is not the case in New Zealand (0.4%).

The share of R&D in agricultural support between 1986-88 and 2007-09 decreased in several countries such as Iceland, Mexico and Turkey. Among the non-OECD countries included in PSE database, South Africa and Brazil had a high share of R&D expenditure in agricultural support in the mid-1990s, but this share declined significantly over time, although it is still higher than the OECD average.

Private expenditure for agricultural R&D

The private sector has always played a role in the development of embodied technologies where knowledge is incorporated into farm inputs such as machinery. For example, the private sector has become more involved in plant breeding as genetic technologies, intellectual property right protection and hybridisation, have allowed private firms to capture a large share of the value of their innovations. Although the private sector is becoming an important supplier of agricultural R&D in many countries, there is only limited information exists on the private expenditure for agricultural R&D. Table 4.2 presents the annual growth rate of private expenditure on agricultural R&D in some OECD countries in the past three decades: 1980s, 1990s and 2000s. In Australia, Iceland, Korea and Spain, the annual growth rate of private R&D expenditure on agriculture had increased significantly in 2000s. In Korea and Spain, the annual growth rate of private R&D expenditures for agriculture far exceeded that that of public R&D expenditure in 2000s. In Australia and Iceland, the growth of private R&D expenditure partly compensated the decline of public R&D expenditures. Private R&D expenditure on agriculture as a share of agricultural GDP is increasing in Australia, Iceland, Korea and Spain (Figure 4.5).

Figure 4.5. Evolution of private expenditure on agricultural R&D in selected OECD countries, 1992, 2000 and 2006

Percentage of agricultural GDP

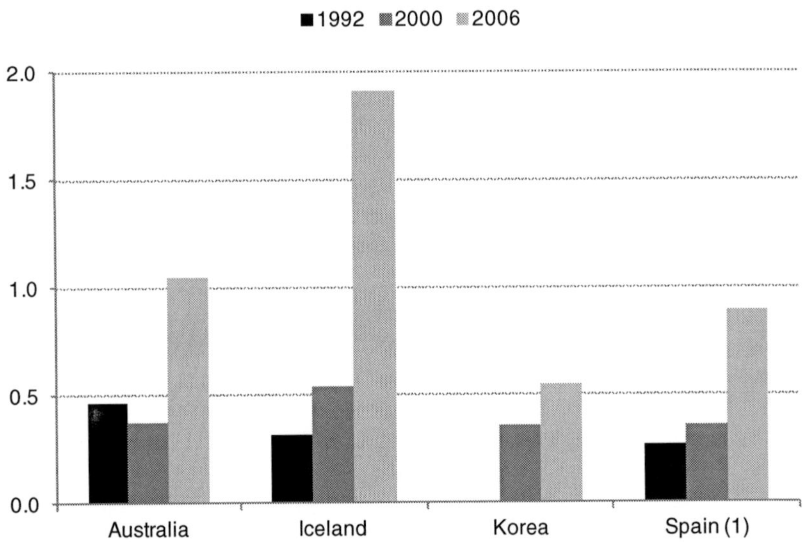

1. Average of 1999 and 2001 instead of 2000.

Source: OECD Research and Development Database.

In some countries, higher educational institutions play an important role in agricultural R&D. In Australia, Korea and Spain, R&D expenditure by such institutions is generally equivalent to that of business enterprises. In general, expenditure on agricultural R&D undertaken by government, business enterprises and higher education institutions increased in Korea and Spain in the 2000s.

Table 4.3. Evolution of expenditure on agricultural R&D by higher education institutions, 1992 to 2008

Country	1980s		1990s		2000s	
	Period	Annual growth rate (%)[1]	Period	Annual growth rate (%)[1]	Period	Annual growth rate (%)[1]
Australia	1981-90	-2.7	1990-00	8.7	2000-08	3.7
Austria	1981-89	2.1	1989-98	15.6	1998-2007	-4.0
Denmark	1982-89	11.9	1990-99	4.6	2000-07	13.7
Iceland	1983-90	18.5	1991-99	-6.2	2000-08	9.4
Ireland	1981-89	14.6	1990-94	20.6		n.a.
Korea		n.a.	1995-00	8.5	2001-08	14.0
Norway	1981-89	5.2	1989-99	-0.5	1999-2003	2.1
Portugal	1982-89	28.5	1990-99	9.9	2000-08	-1.5
Spain		n.a.	1995-99	15.0	2000-08	6.8

n.a.: Not available.

1. Calculated from expenditures in USD Year 2000.

Source: Secretariat's calculations based on OECD R&D Data.

Notes

1. Public expenditure on agricultural extension services are reported in the OECD Producer Support Estimate (PSE) database, available at: www.oecd.org/agriculture/pse.

2. Available at *www.oecd-ilibrary.org/science-and-technology/data/oecd-science-technology-and-r-d-statistics_strd-data-en*.

3. This information is gathered through a questionnaire (common to OECD and Eurostat) sent to member countries' statistical agencies. Data on R&D expenditure generally comes from either specific surveys or administrative data.

4. Eurostat's Nomenclature for the Analysis and Comparison of Scientific Programmes and Budgets 2007 classification.

5. Public funding is based on data from two sources: the National Science Foundation's Federal Funds for Research and Development series (Federal level) and USDA's Current Research Information Systems (State-level). Private funding estimates are constructed by the Economic Research Service (ERS) of the USDA. The data is available on ERS website at: *www.ers.usda.gov/Data/AgResearchFunding/*.

6. OECD data on R&D expenditures are expected to feed into the ASTI database.

7. In Australia, real expenditure on public agricultural R&D reported in the OECD R&D database decreased from USD 469 million in 2000 to USD 231 million (constant 2000 USD) in 2008, while similar expenditure by private enterprises increased from USD 72 million to USD 154 million in the same period in constant 2000 USD. However, Sheng, Gray and Mullen (2011) state that real public R&D expenditure increased in the long term from USD 140 million to USD 829 million (constant 2008 USD) between 1953 and 2007. They also show that growth in public R&D expenditure has slowed down since the 1970s and that the amount is lower in 2007 than in the early 2000s.

8. Available on the OECD website at: *www.oecd.org/agriculture/pse*.

9. The PSE database also contains information on public expenditures on agricultural extension and advisory services, and on agriculture-specific training and education. A project reviewing the coverage and classification of the General Services Support Estimate (GSSE) is expected to lead to improvements in the specification of these expenditures.

Chapter 5

Determinants of productivity growth and competitiveness

A number of studies on agricultural productivity and competitiveness have tried to identify their main determinants. This chapter discusses the results they found regarding the impact of farm size, factor intensity, farm specialisation, human capital, consumer demand, the natural environment, investments in general infrastructures, regulations, and agricultural policies. The impact of R&D on productivity growth is discussed on the basis of the analysis contained in OECD Agricultural Working Paper No. 31 on the impact of R&D investments on productivity growth in agriculture. Estimation issues are first discussed and the importance of a good specification of the lags between investments and their observed benefits is outlined. According to the meta-analysis of over 1 000 estimates of returns to agricultural R&D reported in the Working Paper, the rate of return appears to be quite large, ranging between 20% and 80% per annum in most cases

Review of the impact of possible determinants

To further understand cross-country disparity in productivity and competitiveness and their change over time, many empirical studies have been carried out. Some studies reported in Latruffe (2010) attempt to identify the determinants of productivity and competitiveness, regressing scores over a set of explanatory variable, looking at correlations or using cluster analysis to extract farmers' performance by groups. The comparison of productivity and competitiveness can also be made directly across groups with different characteristics. Other studies specifically investigate the impact of one specific element, for example expenditures in research R&D, on productivity growth. The first section of this chapter considers this specific issue on the basis of results reported in Alston (2010).

Latruffe (2010) distinguishes determinants that are under the managers' control from those that are beyond the managers' control. The first category includes the size of the business, its legal status, factor intensity, product specialisation, production and marketing practices, structure of the land, labour and capital (rented/own), and the characteristics of farm labour. The second category includes factor endowment such as climatic and geographical conditions, general resources in land, labour and capital, consumer demand, government intervention in the agricultural sector (e.g. agricultural policies, regulations, taxation), expenditures in research, extension and infrastructure and location of activities.

The relationship between **farm size** and competitiveness is a widely debated issue, particularly in relation to structural change. A wide range of results is found, depending on the circumstances, the farm type, the type of indicators of size and competitiveness chosen, and the criteria used to define small(er) or large(r) farms. A general finding from these studies is that larger farms are better performers as they can achieve economies of scale and benefit from access to output and input markets (Annex Table A.9). In particular, they suffer less from hidden unemployment. However, there are also some other studies showing that smaller farms are better performers. The main argument to explain this inverse relationship is that very large farms using hired labour may be affected by labour supervision and organisational problems, while family labour is highly motivated as it benefits directly from farm profits. In addition, smaller family farms are also considered to be more resilient because family labour is more adaptable and because family farms are less dependent on external capital than larger farms. Finally, some studies also find that the relationship between farm size and performance is U-shaped, or depend on the farm size variable. It is also obvious that the "optimal" farm size usually depends on the type of production. In the food processing sector, size is less of an issue although smaller firms may be constrained in adopting labour-intensive technologies and face higher input prices.

There is no clear picture in the literature on efficiency superiority of either family farms or corporate farms in OECD and transition countries. Similarly, there is no clear relationship between technical efficiency and **factor intensity** indicators such as capital-labour ratio, or land- labour ratio. Conflicting results are also found regarding the relationship between technical efficiency and the share of hired labour and rented land in, respectively, total labour and total land use. Hired labour may imply better educated workers or workers with specific skills, but may result in supervision problems. Renting land may give farmers an incentive to be productive in order to pay rent, but may prevent them from applying long-term improvements. Regarding the level of indebtedness, some researchers report that this has a positive impact on technical efficiency, suggesting that farmers who are indebted need to meet their repayment obligations and, therefore, are motivated to improve their efficiency. However, highly indebted farmers might incur high credit costs and thus less technically efficiency. In terms of productivity change, borrowing may help farmers to invest in new technology, as found by Zhengfei and Oude Lansink (2006) for Dutch farms for the period 1990-99.

Farm specialisation might be beneficial to technical efficiency since it enables farmers to concentrate their attention on a few tasks and their capital on specific technology, and thereby improve management practices. On the other hand, diversification may improve efficiency by reducing the risk related to the loss of all crops to disease (Latruffe, 2010). Diversification can also result in economies of scope leading to higher efficiency, when several outputs are jointly produced at a lower cost.

The impact of **human capital** on farm technical efficiency and productivity change is often investigated using indicators such as farmers' age or number of years of experience, education level or type, gender, and time spent on the farm. The impact of a farmer's age on technical efficiency can be positive or negative as found in various studies reported by Latruffe (2010). While older farmers may be reluctant or unable to adopt technological innovations, they are more experienced and can use their knowledge to use inputs more efficiently. As expected, most studies found that education has a positive effect on technical efficiency since better educated farm managers are expected to have more skills to run their farm efficiently. Gender is usually not found to affect technical efficiency, although in some developing countries, women might have lower access to inputs. The effect of time spent on off-farm work on performance is ambiguous. While farm managers working off the farm may have less time for managerial activities that would improve farm efficiency work, they might be better able to acquire information and knowledge. Some studies find that part-time farming decreases technical efficiency, others find the opposite (part-time farmers are more efficient) or insignificant results.

When explaining changes in competitiveness or productivity in a sector or a country, some authors mention **consumer demand** (Venturini and Boccaletti, 1998); Viaene and Gellynck, 1998 or Banterle and Carraresi, 2007). As Porter (1990) underlines, the presence of sophisticated and demanding buyers is important in creating and sustaining competitive advantage.

Differences in competitiveness across farms may be explained by the characteristics of the **natural environment** in which they operate (e.g. climate, soil quality, altitude or slope). These are often represented using location dummy variables for regions. They are usually found to have a significant impact on technical efficiency. For example, high quality soils are associated with high technical efficiency. Climate and climatic events are also important. Alston *et al.* (2010) mention catastrophic climatic events to explain bad performance in some years, and increase in population to explain the decrease of labour productivity in some countries.

Higher **density** of a farm type in a region is found to have a positive impact on the technical efficiency in that sector, suggesting knowledge spillovers. Better access to infrastructure and upstream and downstream **facilities** are associated with higher farm technical efficiency. **Public investments in infrastructures** are found to have a positive impact on productivity growth in agriculture, in particular when investment is in public transportation (Ahearn *et al.*, 1998; Yee *et al.*, 2004; Rao *et al.*, 2004), as well as in the food processing industry by acting as a substitute for technological change (Bernstein and Mamuneas, 2008).

Public policies and regulations influence producers' decisions on resource allocation. They may also distort firms' competition (OECD, 2001) and have an effect on competitiveness. Several studies have included a policy indicator in the list of variables used to explain farm competitiveness (Annex Table A.10). They generally find there is a negative correlation between protection and support and competitiveness. The relationship between support and technical efficiency is almost consistently negative across the literature. However, there are diverse results regarding the link between support, and productivity and technological change. Some find a negative correlation, other no significant correlation, while others estimate a positive correlation. For example, support may have a positive effect on technological change as extra income might help farmers overcome their credit constraints and invest in new technology, but the effect on the component efficiency change is not straightforward (Serra *et al.*, 2008). Sauer and Park (2009) report the positive influence of organic subsidies on technical efficiency change and technological change for organic dairy farms in Denmark in the period 2002-04.

The link between government programmes and regulations other than income support, and farm technical efficiency has also been explored. For example, Makki *et al.* (1999) finds that government programme encouraging the diversion of acres from production and conservation reserve programmes had both a negative effect on US agricultural TFP during the period 1930-1990. For German dairy farms during 1987-94 and Greek farms during 1993-97 respectively, Brümmer and Loy (2000) and Rezitis *et al.* (2003) conclude that the European farm credit programme decreased participants' technical efficiency. Larue and Latruffe (2009) find that environmental regulations encourage pig farmers to be more efficient, but that this effect may be counteracted when legal dispositions are too stringent (i.e. when farmers are forced to spread their manure outside their sub-county).

Regarding the **agri-food industry**, regional capital subsidies seem to have had a negative impact on the technical efficiency of food and beverage manufacturing firms in Greece in the period 1989-94 (Skuras *et al.*, 2006). Analysing the effect of trade liberalisation on TFP of five food processing industries in 34 countries (developed and developing) with annual data during 1993-2000, Ruan and Gopinath (2008) conclude that a greater exposure to trade increases productivity, a process that is faster in low productivity countries than in high productivity countries. According to Alpay *et al.* (2002), environmental regulations during 1962-94 were found to have a negative impact on the productivity growth of the Mexican agri-food sector during 1971-94, but not on that in the United States. Based on an opinion survey of 63 stakeholders in the food industry, Wijnands *et al.* (2008) conclude that EU regulation in the sector (which they claim is the third most regulative after the automotive and the chemical sectors) is not a strong obstacle to the competitiveness of the EU15 food sector.

Over the past half century, hundreds of studies have attempted to estimate the impact of agricultural **research and development** on agricultural productivity growth. Main findings are discussed in the following chapter on the basis of a report prepared to that specific end by Julian Alston for the OECD Secretariat (Alston, 2010). The impact of research and development on productivity growth and its economic benefit

The impact of research and development on productivity growth and its economic benefits

Research involves an expenditure of effort to increase the stock of knowledge. Ideas are likely to be generated from the larger stock of knowledge and can be converted (through commercialisation and adoption) into production technologies, good and services, and other forms of innovation. The use of public resources to support agricultural research and development (R&D) begs a critical question of how expenditures on R&D affect the long run competitiveness of the sector.

Estimation issues of R&D impacts on productivity

While there is a good deal of evidence that links R&D expenditure to agricultural productivity growth, quantifying the relationships is challenging and is subject to the availability of suitable methodology and data (Alston, 2010).[1] Some of these data and measurement issues are related to the measurement of R&D expenditure and TFP, while some are related to the often complex problem of relating productivity growth to R&D expenditure. Moreover, measures of agricultural inputs (especially capital, but also farm labour) and outputs are sometimes found problematic. Unpaid owner and family labour, is often an approximation of labour at best. Changes in soil quality or changes in the use of ground water, or influences of changing climate can have important impact but are seldom measured carefully. Limitations on the types and quantities of data that are available, combined with some misunderstanding or misuses of the measures, are likely to have contributed to weaknesses in some studies linking agricultural R&D to productivity.

There are also many data issues in measuring R&D effort. Data on private research expenditures is particularly difficult to obtain as firms often protect this strategic information.[2] Even finding data on public research expenditures in a useful form is often an arduous task because of how agricultural research expenditures are recorded over time. In addition, the data issues with respect to TFP and agricultural R&D expenditures are also confounded by the need to find a series of data which are long enough to reliably estimate the long lags involved in agricultural research particularly in countries which have conducted basic research for extended time periods.

Besides the data issues, attribution problems have bedevilled studies of the effects of research on agricultural productivity (Alston and Pardey, 2001). The principle areas of difficulty are: 1) in determining how much productivity growth is attributable to organised R&D; 2) in attributing responsibility among alternative public and private providers of R&D, and 3) in identifying the research lag structure. Many studies assume implicitly or explicitly that all measured agricultural productivity growth is attributable to R&D. This implicitly assumes that other important drivers of productivity such as education, or infrastructure development, scale economies, clustering effects[3] and changing weather patterns would not have increased productivity growth in the absence of the R&D expenditure. There is also an implicit assumption that productivity would not have decreased due to disease and pest pressure, weather changes or resource depletion in the presence of R&D.

Research usually takes a long time to affect production, and then it affects production for a long time. One element of the attribution problem, then, is to identify the specific dynamic structure linking research spending, knowledge stocks accumulation, and productivity growth. A large number of previous studies have regressed a measure of agricultural output or productivity and variables representing agricultural research and extension, often with a view to estimating the rate of return to research.[4] The specification of the determinants of the lag relationship between research investments and production, which involves the dynamics of knowledge creation, depreciation, and utilisation, is crucial. Only a few studies have presented much in the way of formal theoretical justification for the particular lag models they have employed in modelling returns to agricultural research.

Table 5.1 summarises some key features of research lag distribution models applied in studies of agricultural productivity in OECD countries. Until quite recently, it was common to restrict the lag length to be less than 20 years. In the earliest studies, available time series were short and lag lengths were very short, but the more recent studies have tended to use longer lags. Since the time span of the data set is usually not much longer than the assumed maximum lag length, and the individual lag parameter estimates are unstable and imprecise, most studies have restricted the lag distribution to be represented by a small number of parameters.[5]

Table 5.1. Research lag structures in studies of agricultural productivity

Characteristic	Number of estimates	Estimation period: 1958-69	1970-79	1980-89	1990-98	1958-98
	Count	*Percentage*				
Research lag length (benefits)						
0 to 10 years	253	9.7	6.2	17.9	12.7	13.4
11 to 20 years	537	41.9	22	38.8	22.8	28.5
21 to 30 years	376	0	20.7	12	25.9	19.9
31 to 40 years	178	0	4.3	5.6	14.3	9.4
40 up to ∞ years	141	0	9.5	6.6	7.6	7.5
∞ years	102	35.5	7.5	2.9	5.4	5.4
Unspecified[1]	109	12.9	13.1	3.2	4.9	5.8
Unclear[2]	190	0	16.7	12.7	6.3	10.1
Total	1 886	100	100	100	100	100

Note: This table is based on the full sample of 292 publications reporting 1 886 observations.
1. Unspecified estimates are those for which the research lag length is not made explicit.
2. Lag length is unclear.
Source: Alston *et al.* (2009b), as adapted from Alston *et al.* (2000).

In their application using long-run, state-level data on US agriculture, Alston *et al.* (2009a) found evidence in favour of a gamma lag distribution model with a much longer research lag than most previous studies had found — for both theoretical and empirical reasons.[6] Their empirical work supported a research lag of at least 35 years and up to 50 years for US agricultural research, with a highest correlation in year 24.[7] This comparatively long lag has implications both for econometric estimates of the effects of public R&D on productivity and the implied rate of return to research. It should be noted, however, that lags are likely to depend on the type of research (general or applied, scientific or organisational, by sector, etc.) and the starting point. For example, basic research most likely takes more time to affect productivity gains than applied or adaptive research. Research lags are likely to be longer in OECD countries, which spent significant resource on basic research, than in developing countries, which adopt or adapt existing technologies from international research centres or other countries.

More recently, agricultural economists have been paying increasing attention to the fact that knowledge created within a particular geopolitical entity can have impacts on technology elsewhere, with implications that may matter to both the creators of the spillouts and the recipients of the spillins. For example, Huffman and Evenson (1993) and Alston *et al.* (2010) found that a sizable share of the benefits from research conducted in US State Agricultural Experiment Stations was earned as interstate spillovers. Given the size of these spillovers studies that did not allow for spillovers probably have overestimated the local benefits of research, while underestimating the regional benefits.

Studies that have examined research spillovers have found that knowledge created in neighbouring jurisdictions, or in similar agro-climatic regions can have large impacts on productivity (e.g. Huffman and Evenson, 1993; Pardey *et al.*, 1996 and Alston *et al.*, 2010). Similarly, the varieties and germplasm created in the international research institutions find their way into varieties around the world. Upstream basic research or downstream expenditures on extension can also impose the spillover impacts. Finally private and public research can create spillovers across organisational boundaries and can not only affect research outcome but can also affect research investment decisions by "crowding out" or "crowding in" other research activities. Being able to estimate the spillover effects requires that expenditure data be collected from each potential source of spillover, which further compounds the difficulty of data collection.

Economic benefit of agricultural R&D

Policy makers are fundamentally interested in how investment in R&D affects productivity growth, and whether these investments have a high rate of return relative to the cost capital. Over the past half century or so, hundreds of studies have been published reporting measures of agricultural productivity, the effects of agricultural R&D on agricultural innovation and productivity patterns, and the resulting social payoffs to investments in agricultural R&D. In the standard model of research benefits, research causes the commodity supply curve to shift down and out against a stationary demand curve, giving rise to an increase in quantity produced and consumed, and a lower price (Alston, Norton, and Pardey, 1995). The benefits are assessed using Marshallian measures of research-induced changes in consumer surplus for consumer benefits and of research-induced changes in producer surplus for producer benefits. The total gross annual research benefits depend primarily on the size of the research-induced supply shift and the scale of the industry to which it applies.[8] Other aspects of the analysis typically have second-order effects on the measures of total benefits but may have important implications for the distribution of the benefits between producers and consumers and others.[9]

Measures of the size and distribution of research benefits will be affected by various complications that can be introduced to extend the basic model. Models of research benefits have been extended to incorporate various types of market distortions, including 1) those resulting from the introduction of distortions associated with government policies such as farm commodity programmes or trade barriers (e.g. Alston, Edwards, and Freebairn, 1988); 2) those resulting from the exercise of market power by middlemen (e.g. Huang and Sexton 1996); and 3) those resulting from environmental externalities (e.g. Antle and Pingali 1994). A general finding is that the main effect of a market distortion in this context is to change the distribution of research benefits, with comparatively small effects on the total benefits.

There are mainly two alternatives to assess the economic benefit of agricultural R&D. First, the net present value (NPV) of a stream of research benefits is a widely accepted measurement. This index can be calculated as the difference between the present value of research benefit and cost. In some case, the benefit cost ratio is calculated as a ratio between the present value of research benefits and costs. Second, calculating the internal rate of return (IRR) of the research benefit is also a common way of estimating the benefit of agricultural R&D. IRR is defined as a discount rate that yields NPV equal to zero.

Alston *et al.* (2010) showed that, although the specific estimates were somewhat sensitive to the modelling choice, the annual value of agricultural productivity gains is worth many times more than the annual value of expenditures on research. Consequently the benefits from productivity growth attributed to agricultural R&D exceed the costs by an order of magnitude (i.e. a factor of 10 or more), regardless of methods of measurement or assumptions about attribution (e.g. the shape and length of the R&D lag distribution, inter-regional or inter-institutional spillovers, or the roles of private R&D or extension).

Alston *et al.* (2000a) conducted a comprehensive meta-analysis of studies that had reported estimates of returns to agricultural R&D. The study sample includes 292 studies that reported a total of 1 852 estimates of rates of return to agricultural R&D, from which Alston *et al.* (2000a) reported an overall mean internal rate of return of 81.3%, with a mode of 40%, and a median of 44.3% (Table 5.2). After dropping some outliers and incomplete observations, they conducted regression analysis using a sample of 1 128 estimates with a mean of 64.6%, a mode of 28%, and a median of 42.0%. They found results that were generally consistent with expectations but in many cases they could not distinguish statistically significant effects on the estimated rates of return associated with the nature of the research being evaluated, the industry to which it applied, or the evaluation methodology, because the signal-to-noise ratio was too low. Nevertheless, a predominant and persistent finding across the studies was that the rate of return was quite large. The main mass of the distribution of internal rates of return reported in the literature is between 20% and 80% per annum. Other reviews of the literature may not have covered the same studies or in the same ways, but nevertheless reached similar general conclusions – for example, Evenson (2002), and Fuglie and Heisey (2007). However, Alston *et al.* (2000a) raised a number of concerns about the methods used in the studies that were likely to have led to upwards biases in the estimates. In particular, they suggested that many of the studies may have suffered from estimation bias associated with 1) using research lag distributions that were too short (the results showed that increasing the research lag length resulted in smaller rates of return, as theory would predict), 2) "cherry picking" bias in which only the most successful research investments were evaluated, 3) attribution biases associated with failing to account for the spillover roles of other private and public research agencies, both at home and in other states or other countries, in contributing to the measured benefits, or 4) other aspects of the methods used.

Table 5.2. Lag structures and rates of return to agricultural R&D

Characteristic	Estimates		Rate of return				
	Number	Share of total	Mean	Mode	Median	Minimum	Maximum
	Count		*Percentage*				
Research lag length							
0 to 10	370	20.9	90.7	58	56	-56.6	1 219.0
11 to 20	490	27.7	58.5	49	43.7	-100	677
21 to 30	358	20.2	152.4	57	53.9	0	5 645.0
31 to 40	152	8.6	64	40	41.1	0	384.4
40 to ∞ years	113	6.4	29.3	20	19	0.3	301
∞ Years	57	3.2	49.9	20	35	-14.9	260
Unspecified	205	11.6	48.7	25	34.5	1.1	337
Unclear	27	1.5	43.1	27 and 60	38	9	125
Research gestation lag							
Included	468	59.2	65.5	46	47.1	-14.9	526
Omitted	314	39.7	96.7	95	58.8	0	1 219.0
Unspecified or unclea	8	1	25.1		24.1	6.9	55
Total	790	100	77.5	46 and 58	50.2	-14.9	1 219.0
Spillovers							
Spillins	291	16.7	94.5	95	68	0	729.7
Spillouts	70	4	73.7	95	46.4	8.9	384.4
No spillovers	1 428	81.7	78.8	49 and 57	40	-100	5 645.0

This table is based on a full sample of 292 publications reporting 1 886 observations. For all characteristics, the sample excludes two extreme outliers and includes returns to research only and combines research and extension so that the maximum sample size is 1 772. For the research gestation lag, the sample includes only observations with an explicit lag shape, resulting in a sample size of 790 observations. For spillovers, 25 observations were lost owing to incomplete information, resulting in a sample size of 1 747 observations. Some estimates have spillover effects in both directions.

Source: As reported by Alston *et al.* (2009b), based on data reported in Alston *et al.* (2000a).

Notes

1. This chapter is almost entirely drawn from Alston (2010).

2. As shown in Table 4.2, only six OECD countries provide data on private expenditure on agricultural R&D to the OECD. They are based on national surveys of R&D expenditure.

3. A farmer is more inclined to adopt innovations if its neighbours do so.

4. A comprehensive reporting and evaluation of this literature is provided by Alston *et al.* (2000); see also Schuh and Tollini (1979), Norton and Davis (1981), Evenson (2002) and Alston, Andersen, James and Pardey (2009a).

5. As documented by Alston *et al.* (2000a), common types of lag structures used to construct a research stock include the de Leeuw or inverted-V (e.g. Evenson 1967), polynomial (e.g. Davis 1980; Leiby and Adams 2002; Thirtle and Bottomley 1988), and trapezoidal (e.g. Huffman and Evenson, 1989, 1992, 1993, 2006; Evenson 1996). A small number of studies have used free-form lags (e.g. Ravenscraft and Scherer 1982; Pardey and Craig 1989; Chavas and Cox 1992).

6. The detailed arguments are laid out in Alston, Norton, and Pardey (1995) and some earlier evidence is presented by Pardey and Craig (1988) and Alston, Craig, and Pardey (1998). See also Huffman and Evenson (1989). Alston, Craig, and Pardey (1998) discussed the issue of knowledge depreciation drawing on the previous literature and these arguments are restated and refined by Alston, Pardey, and Ruttan (2008), and Alston, Andersen, James and Pardey (2009a).

7. Alston, Pardey, and Ruttan (2008) documented the adoption lags for particular agricultural technologies and their results are consistent with relatively long overall lags.

8. As noted by Alston, Norton, and Pardey (1995, pp. 60-61), and more recently elaborated by Oehmke and Crawford (2002), the elasticity of supply can have important implications for

measures of research benefits if it is used to translate an assumed horizontal shift into a vertical shift, or vice versa.

9. The distribution of the benefits between producers and consumers depends on the relative elasticities of supply and demand, the nature of the research-induced supply shift and, less importantly, on the functional forms of supply and demand (Alston, Norton, and Pardey, 1995). The nature of the research-induced supply shift has been controversial because it matters, especially for findings concerning the distribution of benefits, and is not easy to observe. Another issue is the distribution of producer benefits among producers. Even if we can be assured that producers as a whole would benefit, those who do not adopt the new technology will not gain and may even be made worse off if the adoption by others leads to price reductions.

Chapter 6

Towards an innovation systems approach

This chapter outlines where more work and data would be needed to better understand agricultural productivity growth and competitiveness in the agricultural and agri-food sector, and the role of R&D. It suggests an "innovation systems" approach would help understand better how innovation translates into productivity growth.

This report provides an overview of conceptual linkages between agricultural competitiveness, productivity growth, technical efficiency change, technological change, innovation and R&D and discusses measurement issues. Latruffe (2010) has reviewed a large range of competitiveness and productivity indicators and methods of estimation. Most authors recognise that a single indicator is not sufficient to assess the broad issue of competitiveness. Some have attempted to created trade-related composite indicators. Still, efforts to compare different methods are limited. Moreover, there is no consensus on how to measure competitiveness, components of competitiveness and its determinants or drivers. This report takes the view that productivity can be used as a good indicator of competitiveness, and reviews the literature and existing data on the cross-country differences in productivity growth and the related role of R&D. There are, however, other indicators that could assess the performance of the sector (e.g. profitability, environmental performance) and other determinants of productivity growth, which are briefly reviewed. In particular, the non-price components of competitiveness, such as product differentiation, product and service quality and variety, design, novelty, reputation, reliability and sustainability, are usually excluded from the analysis, probably because they are difficult to measure.

While evidence on productivity growth and competitiveness has been for a long time relatively scattered, recent studies have estimated developments in productivity growth on a larger scale. Alston *et al.* (2010) represent a major effort to measure trends in productivity growth across a number of large producing countries and at regional and global level, using different methods and indicators of partial and total factor productivity. Ball *et al.* (2001, 2006 and 2010) also represent a consolidated effort to compare developments in EU member states and the United States, using comparable datasets and indicators. In the same vein, Butault and Réquillart (2010) examine a longer and more recent period (1959-2008). Still, comparative studies of productivity are restricted in their country, time and indicator coverage by availability of comparable and reliable data. In the European Union, studies have focussed on estimating productivity and its components at the farm level. Because of the availability of farm level data, they mainly cover the 1990s and at most the early 2000s. They often include comparisons between EU member states and major partners. While several countries monitor agricultural productivity developments on a regular basis and some have investigated recent developments in farm productivity across farm types, there is no systematic and regular monitoring of productivity growth and competitiveness in the agri-food sector on a comparable basis at international level.

Evidence from a wide range of literature suggests the central role of R&D in fostering agricultural productivity growth. It also suggests that the benefits have been worth many times more than the costs. However, results depend on the data and the specification of the model used for measuring research benefits. Analysis has revealed some areas where findings are sensitive to modelling choices, including the representation of technological change in the model, the treatment of spillovers, and the R&D lag distribution. Other specification choices, such as how to deal with distortions from market power of firms, government policy, or environmental externalities, have been shown to have relatively important effects on estimates of the distribution of benefits and relatively little effect on estimates of the size of total benefits. The creation of the "data" used in analyses is a critical step since the interpretation of results often depends crucially on the data. Some authors point to the need for long time series to estimate TFP in order to better assess lagged responses. Moreover, data on public and private R&D expenditure by region and type of expenditures would help assess the type of research that is most effective in fostering productivity improvement. It may also help understand the characteristics and causes of diffusion lags of research with different types (general, technical or organisational) and purposes (crop, livestock or environment).

This report provides some evidence on agricultural R&D expenditures collected by the OECD and the ASTI. The information contained in these databases is generally collected at the national level, covering the whole agricultural sector in the PSE database and agricultural, hunting, forestry and fisheries in the ASTI and OECD R&D databases. The accuracy and consistency of public R&D expenditures on agriculture in the PSE database over time and across countries are currently being reviewed as well as public expenditures on agricultural schools and technical assistance. The OECD R&D database contains some information on public and private R&D expenditures on agriculture, hunting, forestry and fisheries in OECD countries, while the ASTI database has the same information for a number of emerging and developing economies. It should be noted however, that R&D expenditures are measuring efforts and not achievements.

Many factors other than R&D affect agricultural productivity growth. They have also been investigated in the literature but to a lesser extent and with less consistent results. Moreover, there is little evidence on productivity and competitiveness in the agri-food industry and its determinants. The issue of government intervention could be given more attention. Efforts so far have focussed on the impact of public R&D expenditures on agricultural productivity growth and to some extent the impact of agricultural support policies on productivity and competitiveness. There is little evidence on the impact of specific agricultural policies, such as support to farm investment, specific production practices or marketing facilities. Public support on extension services and infrastructure is expected to have a strong effect, in particular in developing countries. In addition, the impact of other types of interventions such as environmental, labour and fiscal policies on the competitiveness of the agri-food sector has not been assessed.

While formal R&D is central to innovation, it is increasingly recognised that it is not the only source of discovering new technologies for farmers and others. Many new technologies are created without basic science underpinning. More recently, the interactive relationships among basic science, applied science and technology development are emphasised (OECD, 2009). This suggests that a more comprehensive analytical framework going beyond the linear relationship between R&D expenditure and productivity growth[1] would need to be adopted to analyse "innovation systems" in agriculture. The latest edition of the Oslo Manual defines innovation as the introduction of new or significantly improved goods and services, or the use of new inputs, processes, organisational or marketing methods (OECD and Eurostat, 2005). The process of innovation and productivity growth includes not only knowledge creation, but also the whole system of technological diffusion, adoption processes, interactions and market adjustments.

The theory of induced innovation, suggests that shifts in end-user demand can drive the innovation process by creating demand for specific forms of technology. This relationship is especially important for applied research where the firm has to make investments to develop and commercialise technologies. The theory suggests that the direction of technological progress is also determined by market demand. For example, the lack of final demand due to a market failure, (e.g. the lack of a carbon market) can impede the commercialisation and adoption of new technologies (i.e. carbon mitigation technologies). The process of innovation is not as simple as the exogenous invention of new technology or knowledge through R&D.

Increasingly "innovation systems" are viewed as a network of knowledge flows with considerable two way flows of information upstream and downstream and spillovers of knowledge among the participants that are connected in formal and informal ways. This more systemic approach suggests that innovation policy goes far beyond research expenditures and involves a wide range of institutions that can affect incentives, knowledge sharing and the processes used for commercialisation.

This review of concepts and evidence on linkages between R&D, productivity growth and competitiveness also points to the need to adopt a more "innovation systems" approach in agriculture. A conceptual framework could be developed as well as multiple indicators that would help assess the performance of each aspect of innovation systems in agriculture across countries.

The evolution of agricultural R&D expenditure shows that diverse patterns of agricultural R&D have emerged in terms of sources of funding and the institutions undertaking research activities. Future work on innovation systems could take a closer look at institutional arrangements in agricultural innovation and knowledge systems, and would examine the respective roles for the government versus the private sector in strengthening innovation systems and facilitating technological adoption, including research collaboration across sectors; protection of intellectual property rights; and knowledge flow. The Conference on Agricultural Knowledge System (AKS), organised by OECD and which took place in June 2011,[2] looked at developments in AKS institutions and the relationship between the different components at the national and international levels, and discussed whether these are functioning well and responsive to emerging issues. It also reviewed incentives and disincentives to both public and private activities in the AKS, and looked at policy coherence and the best practices. This conference provided valuable information on the performance of innovation systems in OECD countries and selected emerging economies, in particular regarding institutional, regulatory and policy aspects. It also covered transboundary issues such as Intellectual property rights and technology transfers.

The literature review on competitiveness, productivity and efficiency in the agricultural sector concludes that competitiveness is a relative concept. Firms, sectors and nations should be compared with each other. Farm-level analysis of productivity and competitiveness matters as it shows the dynamism and the diversity within the sector. Farm-level data could be used in future work to assess the pathways of productivity growth; investigate the different rates at which farms adopt new technological innovations; and investigate the relationship between agricultural policies, innovation, adoption and diffusion, and productivity growth.

Notes

1. Earlier models viewed innovation as a linear process or a pipeline model, where discoveries, emanating from basic science, lead to efforts in applied science, which subsequently lead to development, to commercialisation and finally innovation leading to productivity improvement.

2. Information on the OECD AKS Conference is available at: *www.oecd.org/agriculture/policies/innovation*.

References

Ahearn, M., D. Culver, and R. Schoney (1990), "Usefulness and limitations of COP estimates for evaluating international competitiveness: a comparison of Canadian and U.S. wheat", *American Journal of Agricultural Economics*, Vol. 72, No. 5, pp. 1283-1291.

Ahearn, M., J. Yee, E. Ball, and R. Nehring (1998), *Agricultural Productivity in the United States*. Agriculture Information Bulletin No. 740, Resource Economics Division, Economic Research Service, US Department of Agriculture, Washington D.C.

Alpay, E., S. Buccola, and J. Kerkvliet (2002), "Productivity growth and environmental regulation in Mexican and U.S. food manufacturing", *American Journal of Agricultural Economics*, Vol. 84, No. 4, pp. 887-901.

Alston, J.M., G.W. Edwards and J.W. Freebairn (1988), "Market Distortions and the Benefits from Research", *American Journal of Agricultural Economics*, Vol. 70, No. 2 (May), pp. 281-288.

Alston, J.M., G.W. Norton, and P.G. Pardey (1995), *Science Under Scarcity: Principles and Practice for Agricultural Research Evaluation and Priority Setting*, Ithaca, Cornell University Press, 1995 (reprinted in soft cover by CAB International 1998).

Alston, J.M., B.J. Craig, and P.G. Pardey (1998), "Dynamics in the Creation and Depreciation of Knowledge, and the Returns to Research", EPTD Discussion Paper No. 35, Washington D.C., International Food Policy Research Institute, August.

Alston, J.M., C. Chan-Kang, M.C. Marra, P.G. Pardey, and T.J. Wyatt (2000a), *A Meta Analysis of Rates of Return to Agricultural R&D: Ex Pede Herculem?* Washington D.C., IFPRI Research Report No. 113.

Alston, J.M., M.C. Marra, P.G. Pardey, and T.J. Wyatt (2000b), "Research Returns Redux: A Meta Analysis of the Returns to Agricultural R&D", *Australian Journal of Agricultural and Resource Economics*, Vol. 44, No. 2, pp. 185-215.

Alston, J.M. and P.G. Pardey (2001), "Attribution and Other Problems in Assessing the Returns to Agricultural R&D", *Agricultural Economics*, Vol. 25, No. 2-3(September), pp. 141-152.

Alston, J.M., J.M. Beddow, and P.G. Pardey (2008a), "Agricultural Research, Productivity, and Food Commodity Prices", *Agriculture and Resource Economics Update*, Vol. 12, No. 2, November-December, pp. 11-14. Available at: *www.agecon.ucdavis.edu/extension/update/articles/v12n2_5.pdf*

Alston, J.M., P.G. Pardey and V.W. Ruttan (2008b), "Research Lags Revisited: Concepts and Evidence from US Agriculture." Paper presented at the 2008 Economic History Association meetings, New Haven CT, 12-14 September.

Alston, J.M., M.A. Andersen, J.S. James, and P.G. Pardey (2009a), "The Economics of Agricultural R&D", *Annual Review of Resource Economics,* Vol. 1 (October), pp. 15-35.

Alston, J.M., J.M. Beddow, and P.G. Pardey (2009b), "Mendel versus Malthus: Research, Productivity, and Food Prices in the Long Run", InSTePP Working Paper, St Paul, University of Minnesota, January.

Alston, J. (2010), "The Benefits from Agricultural Research and Development, Innovation, and Productivity Growth", *OECD Food, Agriculture and Fisheries Working Papers*, No. 31. doi: *10.1787/5km91nfsnkwg-en*

Alston, J.M., M.A. Andersen, J.S. James, and P.G. Pardey (2010), *Persistence Pays: U.S. Agricultural Productivity Growth and the Benefits from Public R&D Spending*, New York, Springer.

Alston, J.M., B.A. Babcock, and P.G. Pardey (eds.) (2010), *The Shifting Patterns of Agricultural Productivity Worldwide,* CARD-MATRIC Electronic Book, Center for Agricultural and Rural Development, The Midwest Agribusiness Trade Research and Information Center, Iowa State University, Ames, Iowa. Available at: *www.matric.iastate.edu/shifting_patterns.*

Antle, J.M. and P.L. Pingali (1994), "Pesticides, Productivity, and Farmer Health: A Philippines Case Study", *American Journal of Agricultural Economics*, Vol. 76 (August), pp. 418-430.

Ball V.E., J.C. Bureau, R. Nehring, and A. Somwaru (1997), "Agricultural Productivity Revisited", *American Journal of Agricultural Economics*, Vol. 79, No. 4, November, pp. 1045-63.

Ball, E., Bureau, J.-C., Butault, J.-P. and Nehring, R. (2001), "Levels of farm sector productivity: An international comparison", *Journal of Productivity Analysis*, Vol. 15, pp. 5-29.

Ball, E., Butault, J.-P., C. San Juan Mesonada, and R. Mora (2006), *Productivity and International Competitiveness of European Union and United States Agriculture, 1973-(2002),* paper presented at the AIEA2 International Meeting "Competitiveness in agriculture and the food industry: US and EU perspectives", Bologna, June.

Ball, V.E., J.P. Butault, C. San Juan Mesonada, and R. Mora (2010), "Productivity and International competitiveness of agriculture in the European Union and the United States", *Agricultural Economics*, Vol. 41, pp. 611-627.

Bakucs, L., L. Latruffe, I. Fertö and J. Fogarasi (2010), "Impact of EU accession on farms' technical efficiency in Hungary", *Post-Communist Economies*, Vol. 22, No. 2, June, pp. 165-175.

Banse, M., M. Gorton, J. Hartel, G. Hughes, J. Köckler, T. Möllman, and W. Münch (1999), "The evolution of competitiveness in Hungarian agriculture: From transition to accession", *MOCT-MOST*, Vol. 9, pp. 307-318.

Banterle, A. and L. Carraresi (2007), "Competitive performance analysis and European Union trade: The case of the prepared swine meat sector", *Food Economics – Acta Agricult Scand C*, Vol. 4, pp. 159-172.

Barnes, A.P., C. Revoredo-Giha, J. Sauer, J. Elliott, and G. Jones (2010), *A report on technical efficiency at the farm level 1989 to 2008*, Report for Defra, London.

Bavorova, M. (2003), "Influence of policy measures on the competitiveness of the sugar industry in the Czech Republic", *Agricultural Economics – Czech*, Vol. 49, No. 6, pp. 266-274.

Bernstein, J. and T. Mamuneas (2008), "Public infrastructure, input efficiency and productivity growth in the Canadian food processing industry", *Journal of Productivity Analysis*, Vol. 29, pp. 1-13.

Bezlepkina, I., A. Oude Lansink, A. and Oskam (2005), "Effects of subsidies in Russian dairy farming", *Agricultural Economics*, Vol. 33, pp. 277-288.

Block, S. (2010), *The decline and rise of agricultural productivity in Sub-Saharan Africa since 1961*, National Bureau of Economic Research Working Paper 16481, October. Available at: *www.nber.org/papers/w16481*

Bojnec, S. (2003), "Three concepts of competitiveness measures for livestock production in Central and Eastern Europe", *Agriculturae Conspectus Scientificus*, Vol. 68, No. 3, pp. 209-220.

Bojnec, S. and I. Fertö (2009), "Agro-food trade competitiveness of Central European and Balkan countries", *Food Policy*, Vol. 34, pp. 417-425.

Bojnec, S. and L. Latruffe (2007), *Farm Size and Efficiency: The Case of Slovenia*, contributed paper presented at the 100th seminar of the EAAE, “Development of Agriculture and Rural Areas in Central and Eastern Europe”, Novi Sad, Serbia, 21-23 June.

Brümmer, B. and J.-P. Loy (2000), “The technical efficiency impact of farm credit programmes: A case study in Northern Germany”, *Journal of Agricultural Economics*, Vol. 51, No. 3, pp. 405-418.

Brümmer, B, T. Glauben, and G. Thijssen (2002), “Decomposition of productivity growth using distance functions: The case of dairy farms in three European countries”, *American Journal of Agricultural Economics*, Vol. 84, No. 3, pp. 628-644.

Buccola, S., Y. Fujii, and Y. Xia (2000), “Size and productivity in the US milling and baking industries”, *American Journal of Agricultural Economics*, Vol. 82, No. 4, pp. 865-880.

Bureau, J.-C. and J.-P. Butault (1992), "Productivity gaps, price advantages and competitiveness in E.C. agriculture", *European Review of Agricultural Economics*, Vol. 19, No. 1, pp. 25-48.

Bureau, J.-C., J.-P. Butault, and A. Hoque (1992), *International Comparisons of Costs of Wheat Production in the EC and United States*, staff Report No. 9222, Economic Research Service, United States Department of Agriculture, Washington D.C.

Bush, V. (1945), *Science: the Endless Frontier*, US Government Printing Office, Washington D.C.

Butault , J.P. and V. Réquillart (2010), *La stagnation de la production, l'essouflement des gains de productivité et le déclin de la compétitivité de l'agro-alimentaire en France*, Working Paper for MAAPRAT, INRA-SAE2, November.

Chan-Kang, C., S. Buccola, and J. Kerkvliet (1999), “Investment and productivity in Canadian and US food manufacturing”, *Canadian Journal of Agricultural Economics*, Vol. 47, No. 2, pp. 105-118.

Carraresi, L. and A. Banterle (2008), *Measuring Competitiveness in the EU Market: A Comparison Between Food Industry and Agriculture*, paper presented at the 12th EAAE Congress, Gent, Belgium, 27-30 August.

Carroll, J., S. Greene, C. O’Donoghue, C. Newman, and F. Thorne (2009), *Productivity and the Determinants of Efficiency in Irish Agriculture (1996-2006)*, paper presented at the 83rd AES Conference, Dublin, Ireland, 30 March-1 April.

Chavas, J.-P., and T.L. Cox (1992), “A Nonparametric Analysis of the Effects of Research on Agricultural Productivity”, *American Journal of Agricultural Economics*, Vol. 74 (August), pp. 583-591.

Coelli, T., D. Rao, C., O'Donnell, and G. Battese (2005), *An Introduction to Efficiency and Productivity Analysis*, Springer, New York, second edition.

Coelli, T.J. and D.S. Prasada Rao (2005), "Total factor productivity growth in agriculture: A Malmquist index analysis of 93 countries, 1980–2000", *Agricultural Economics*, Vol. 32, pp. 115-134. Available at: *ideas.repec.org/a/bla/agecon/v32y2005is1p115-134.html*

Davidova, S., M. Gorton, B. Iraizoz, and T. Ratinger (2003), "Variations in farm performance in transitional economies: Evidence from the Czech Republic", *Journal of Agricultural Economics*, Vol. 54, No. 2, pp. 227-245.

Davis, J.S. (1980), "A Note on the Use of Alternative Lag Structures for Research Expenditure in Aggregate Production Function Models", *Canadian Journal of Agricultural Economics*, Vol. 28, pp. 72-76.

Diewert, W. E. (1992), "The Measurement of Productivity", *Bulletin of Economic Research*, Blackwell Publishing, Vol. 44, No. 3, pp. 163-98, July.

Drescher, K. and O. Maurer (1999), "Competitiveness of the European dairy industries", *Agribusiness*, Vol. 15, No. 2, pp. 163-177.

Emvalomatis, G., A. Oude Lansink, and S. Stefanou (2008), *An Examination of the Relationship Between Subsidies on Production and Technical Efficiency in Agriculture: The Case of Cotton Producers in Greece*, paper presented at the 107th EAAE Seminar 'Modelling of Agricultural and Rural Development Policies', Seville, Spain, 29 January-1 February.

Evenson, R.E.(1967), "The Contribution of Agricultural Research to Production", *Journal of Farm Economics*, Vol. 49 (December), pp. 1415-1425.

Evenson, R.E. (1996), "Two Blades of Grass: Research for U.S. Agriculture", Chapter 11 in J.M. Antle and D.A. Sumner (eds.), *The Economics of Agriculture Volume 2, Papers in Honor of D. Gale Johnson*, Chicago, University of Chicago Press, pp. 171-203.

Evenson, R.E. (2002),"Economic Impacts of Agricultural Research and Extension", Chapter 11 in B.L. Gardner and G.C. Rausser eds. *Handbook of Agricultural Economics, Volume 1A: Agricultural Production*, New York, Elsevier.

Fan, S., P. Hazell and S. Thorat (1999), "Linkages between Government Spending, Growth and Poverty in Rural India", *Research Report 110*, International Food Policy Research Institute, Washington D.C.

Fertö, I. and L. Hubbard (2003), "Revealed comparative advantage and competitiveness in Hungarian agri-food sectors", *World Economy*, Vol. 26, No. 2, pp. 247-259.

Fischer, C. and S. Schornberg (2007), "Assessing the competitiveness situation of EU food and drink manufacturing industries: an index-based approach", *Agribusiness: An International Journal*, Vol. 23, No. 4, pp. 473-496.

Fogarasi, J. and L. Latruffe (2009), *Farm Performance and Support in Central and Western Europe: A Comparison of Hungary and France*, Working Paper SMART-LERECO No. 09-07, Rennes, France.

Fuglie, K.O. and P.W. Heisey (2007), *Economic Returns to Public Agricultural Research*. USDA, ERS Economic Brief No. 10, Washington D.C., USDA, September.

Galonopoulos, K., Y. Surry, and K. Mattas (2008), *Agricultural Productivity Growth in the Euro-Med Region: Is there Evidence of Convergence?*, paper presented at the 12th EAAE Congress, Gent, Belgium, 27-30 August.

Giannakas, K., R. Schoney, and V. Tzouvelekas (2001), "Technical efficiency, technological change and output growth of wheat farms in Saskatchewan", *Canadian Journal of Agricultural Economics*, Vol. 49, pp. 135-152.

Gopinath, M. (2003), "Cross-country differences in technology: The case of the food processing industry", *Canadian Journal of Agricultural Economics*, Vol. 51, pp. 97-107.

Gopinath, M., C. Arnade, M., Shane, and T. Roe (1997), "Agricultural competitiveness: The case of the United States and major EU countries", *Agricultural Economics*, Vol. 16, pp. 99-109.

Gorton, M., S. Davidova, and T. Ratinger (2000), "The competitiveness of agriculture in Bulgaria and the Czech Republic *vis-à-vis* the European Union", *Comparative Economic Studies*, Vol. 42, No. 1, pp. 59-86.

Gorton, M. and S. Davidova (2001), "The international competitiveness of CEEC agriculture", *World Economy*, Vol. 24, No. 2, pp. 185-200.

Gorton, M., A. Danilowska, S. Jarka, S. Straszewski, A. Zawojska, and E. Majewski (2001), "The international competitiveness of Polish agriculture", *Post-Communist Economies*, Vol. 13, No. 4, pp. 445-457.

Hadley, D. (2006), *Efficiency and Productivity at the Farm Level in England and Wales 1982 to (2002),* report for the Department for Environment, Food and Rural Affairs (DEFRA), London, UK, March.

Hallam, D. and D. Machado (1996), "Efficiency analysis with panel data: A study of Portuguese dairy farms", *European Review of Agricultural Economics*, Vol. 23, pp. 79-93.

Helfand, S. and E. Levine (2004), "Farm size and the determinants of productive efficiency in the Brazilian Center-West", *Agricultural Economics*, Vol. 31, pp. 241-249.

Huang, S.Y. and R.J. Sexton (1996), "Measuring returns to an innovation in an imperfectly competitive market: Application to mechanical harvesting of processing tomatoes in Taiwan", *American Journal of Agricultural Economics,* Vol. 78, pp. 558-571.

Huffman, W.E. and R.E. Evenson (1989), "Supply and Demand Functions for Multiproduct US Cash Grain Farms: Biases Caused by Research and other Policies", *American Journal of Agricultural Economics,* Vol. 71(August), pp. 761-773.

Huffman, W.E. and R.E. Evenson (1992), "Contributions of Public and Private Science and Technology to US Agricultural Productivity", *American Journal of Agricultural Economics,* Vol. 74 (August), pp. 752-756.

Huffman, W.E. and R.E. Evenson (1993), *Science for Agriculture: A Long-Term Perspective,* Ames, Iowa State University Press.

Huffman, W. and Evenson, R. (2001), "Structural and productivity change in US agriculture, 1950-1982", *Agricultural Economics,* Vol. 24, pp. 127-147.

Huffman, W.E. and R.E. Evenson (2006), "Do Formula or Competitive Grant Funds have Greater Impacts on State Agricultural Productivity", *American Journal of Agricultural Economics,* Vol. 88, No. 4, pp. 783-98.

Huffman, W.E. and R.E. Evenson (2006), *Science for Agriculture: A Long-Term Perspective*, Oxford, Blackwell Publishing, second edition.

Krugman, P. R. (1996), "Making sense of the competitiveness debate", *Oxford Review of Economic Policy*, Vol. 12, No. 3, pp. 17-25.

Krugman, P. R. (2001), "Competitiveness: A Dangerous Obsession", *Foreign Affairs*, Vol. 73, No. 2, pp. 28-44.

Lachaal, L. (1994), "Subsidies, endogenous technical efficiency and the measurement of productivity growth", *Journal of Agricultural and Applied Economics*, Vol. 26, No. 1, pp. 299-310.

Larue, S. and L. Latruffe (2009), *Agglomeration Externalities and Technical Efficiency in French Pig Production*, Working Paper SMART-LERECO No. 09-10, Rennes, France.

Latruffe, L., K. Balcombe, S., Davidova, and K. Zawalinska (2004), "Determinants of technical efficiency of crop and livestock farms in Poland", *Applied Economics*, Vol. 36, No. 12, pp. 1255-1263.

Latruffe, L., K. Balcombe, S., Davidova, and K. Zawalinska (2005), "Technical and scale efficiency of crop and livestock farms in Poland: Does specialisation matter?" *Agricultural Economics*, Vol. 32, No. 3, pp. 281-296.

Latruffe, L., K. Balcombe, and S. Davidova (2008), "Productivity change in Polish agriculture: An application of a bootstrap procedure to Malmquist indices", *Post-Communist Economies*, Vol. 20, No. 4, pp. 449-460.

Latruffe, L. (2010), "Competitiveness, Productivity and Efficiency in the Agricultural and Agri-Food Sectors", *OECD Food, Agriculture and Fisheries Working Papers*, No.30. doi: *10.1787/5km91nkdt6d6-en*

Leiby, J.D. and G.D. Adams (2002), "The Returns to Agricultural Research in Maine: The Case of a Small Northeastern Experiment Station", *Northeastern Journal of Agricultural and Resource Economics*, Vol. 20, pp. 1-14.

Liefert, W. (2002), "Comparative (dis?) advantage in Russian agriculture", *American Journal of Agricultural Economics*, Vol. 84, No. 3, pp. 762-767.

Lovell, C. (1993), "Production frontiers and productive efficiency", in Fried, H., C. Lovell, and S. Schmidt (eds), *The Measurement of Productive Efficiency: Techniques and Applications*, Oxford University Press, New York, pp. 3-67.

Makki, S., L. Tweeten, and C. Thraen (1999), "Investing in research and education versus commodity programs: Implications for agricultural productivity", *Journal of Productivity Analysis*, Vol. 12, pp. 77-94.

Mulder, N., A. Vialou, B. David, M. Rodriguez, and M. Castilho (2004), *La compétitivité de l'agriculture et des industries agroalimentaires dans le Mercosur et l'Union Européenne dans une perspective de libéralisation commerciale*, Working Paper/Document de travail N°2004-19, Centre d'Etudes Prospectives et d'Informations Internationales (CEPII), Paris, France, November.

Mullen, J.D. (2010), "Trends in Investment in Agricultural R&D in Australia and its Potential Contribution to Productivity", *Australasian Agribusiness Review*, Vol. 18, Paper 2. Available at: *www.agrifood.info/review/2010/Mullen.pdf*.

Munroe, D. (2001), "Economic efficiency in Polish peasant farming: An international perspective", *Regional Studies*, Vol. 35, No. 2, pp. 461-471.

Nasr, R., P. Barry, and P. Ellinger (1998), "Financial structure and efficiency of grain farms", *Agricultural Finance Review*, Vol. 58, pp. 33-48.

Nivievskyi, O. and S. von Cramon-Taubadel (2008), *The Determinants of Dairy Farming Competitiveness in Ukraine*, paper presented at the 12th EAAE Congress, Gent, Belgium, 27-30 August.

Norton, G.W. and J.S. Davis (1981), "Evaluating returns to agricultural research: A review", *American Journal of Agricultural Economics*, Vol. 63, pp. 685–699.

Nossal, K and Y. Sheng (2010), "Productivity growth: Trends, drivers and opportunities for broadacre and dairy industries", *Australian Commodities*, Vol. 17, pp. 216–230.

OECD (2001), *Competition Policy in Subsidies and State Aid,* OECD, Paris.

OECD and Eurostat (2005), *Oslo Manual: Guidelines for Collecting and Interpreting Innovation Data*, OECD, Paris.

OECD (2009), "Technology and Innovation in World Agriculture: Prospects for 2010-2019" [TAD/CA/APM/CFS/MD/RD(2009)10].

OECD (2010), *OECD Innovation Strategy: Getting a Head Start on Tomorrow*, OECD, Paris.

OECD (2011), *Evaluation of agricultural policy reforms in the European Union*, OECD, Paris.

Oehmke, J.F. and E.W. Crawford (2002), "The Sensitivity of Returns to Research Calculations to Supply Elasticity", *American Journal of Agricultural Economics*, Vol. 64, No. 2, pp. 366-369.

O'Neill, S. and A. Matthews (2001), "Technical efficiency in Irish agriculture", *The Economic and Social Review*, Vol. 32, No. 3, pp. 263-284.

Pardey, P.G. and B. Craig (1989), "Causal Relationships Between Public Sector Agricultural Research Expenditures and Output", *American Journal of Agricultural Economics*, Vol. 71 (February), pp. 9-19.

Pardey, P.G., J. M. Alston, C. Chan-Kang, E. C. Magalhaẽs, and S. A. Vosti (2006), "International and Institutional R&D Spillovers: Attribution of Benefits among Sources for Brazil's New Crop Varieties", *American Journal of Agricultural Economics*, Vol. 88, No. 1 (February), pp. 104–123.

Peterson, E. and S. Valluru (2000), "Agricultural comparative advantage and government policy interventions", *Journal of Agricultural Economics*, Vol. 51, No. 3, pp. 371-387.

Porter, M. E., C. Ketels, and M. Delgado (2007), "The Microeconomic Foundations of Prosperity: Findings from the Business", The Global Competitiveness Report 2007-2008 © 2007 World Economic Forum (*siteresources.worldbank.org/INTRANETTRADE/.../Porter_GCI_ch1_2.pdf*). Accessed 6 December 2010.

Porter, M. (1990), *The Competitive Advantage of Nations*, The Free Press, New York.

Qineti, A., M. Rajcaniova, and E. Matejkova (2009), "The competitiveness and comparative advantage of the Slovak and the EU agri-food trade with Russia and Ukraine", *Agricultural Economics – Czech*, Vol. 55, No. 8, pp. 375-383.

Rao, P., T. Coelli, and M. Alauddin (2004), *Agricultural Productivity Growth, Employment and Poverty in Developing Countries, 1970-(2000),* Employment Strategy Paper 2004/9, Employment Trends Unit, Employment Strategy Department, International Labour Organisation, Geneva, Switzerland.

Ravenscraft, D. and F.M. Scherer (1982), "The Lag Structure of Returns to Research and Development", *Applied Economics*, Vol. 14, pp. 603-620.

Rezitis, A., K. Tsiboukas, and S. Tsoukalas (2003), "Investigation of factors influencing the technical efficiency of agricultural producers participating in farm credit programs: The case of Greece", *Journal of Agricultural and Applied Economics*, Vol. 35, No. 3, pp. 529-541.

Rios, A. and G. Shively (2006), "Farm size and nonparametric efficiency measurements for coffee farms in Vietnam", *Forests, Trees, and Livelihoods*, Vol. 16, pp. 397-412.

Robbins, C.A. and C.E. Moylan (2007) *Research and Development Satellite Account Update: Estimates for 1959-2004*. Washington D.C.: Bureau of Economic Analysis, 2007. Accessed November 2007 from *www.bea.gov/scb/pdf/2007/10%20October/1007_rd_text.pdf.*

Ruan, J. and M. Gopinath (2008), "Global productivity distribution and trade liberalisation: Evidence from processed food industries", *European Review of Agricultural Economics*, Vol. 35, No. 4, pp. 439-460.

Rungsuriyawiboon, S. and A. Lissitsa (2006), "Agricultural productivity growth in the European Union and transition countries", IAMO Discussion Paper No. 94. Available at: *www.nesdb.go.th/econSocial/macro/TNCE/Download/1/supawat.pdf.*

Sauer, J. and T. Park (2009), "Organic farming in Scandinavia: Productivity and market exit", *Ecological Economics*, Vol. 68, pp. 2243-2254.

Schuh, G.E. and H. Tollini (1979), *Costs and Benefits of Agricultural Research: State of the Arts*. World Bank Staff Work Paper No. 360 Washington D.C, World Bank.

Serra, T., D. Zilberman, and J. Gil (2008), "Farms' technical inefficiencies in the presence of government programs", *The Australian Journal of Agricultural and Resource Economics*, Vol. 52, pp. 57-76.

Sharma, K., P. Leung, and H. Zaleski (1999), "Technical, allocative and economic efficiencies in swine production in Hawaii: A comparison of parametric and nonparametric approaches", *Agricultural Economics*, Vol. 20, pp. 23-35.

Sheng, Y., J.D. Mullen, and S. Zhao (2010), "Has Growth in Productivity in Australian Broadacre Agriculture Slowed?" paper presented at the 2010 Conference of the Australian Agricultural and Resource Economics Society (AARES), 10–12 February 2010, Adelaide, South Australia.

Sheng, Y, E.M. Gray, and J.D. Mullen (2011), *Public investment in R&D and extension and productivity in Australian broadacre agriculture,* ABARE Conference Paper 10.16a, prepared for the 16th World Productivity Conference and 2010 European Productivity Conference in Antalya, Turkey, November 2010.

Skuras, D., K. Tsekouras, E., Dimara, and D. Tzelepis (2006), "The effects of regional capital subsidies on productivity growth: A case study of the Greek food and beverage manufacturing industry", *Journal of Regional Science*, Vol. 46, No. 2, pp. 355-381.

Swanson, B.E. and R. Rajalahti (2010), *Strengthening Agricultural Extension and Advisory Systems: Procedures for Assessing, Transforming, and Evaluating Extension Systems*, World Bank Agriculture and Rural Development Discussion Paper 45, Washington D.C.

Thirtle, C.G. and P. Bottomley (1988), "Is Publicly Funded Agricultural Research Excessive?" *Journal of Agricultural Economics*, Vol. 31, pp. 99-111.

Thorne, F. (2005), *Analysis of the Competitiveness of Cereal Production in Selected EU Countries*, paper presented at the 11th EAAE Congress, Copenhagen, Denmark, 24-27 August.

Toming, K. (2007), "The impact of EU accession on the export competitiveness of Estonian food processing industry", *Post-Communist Economies*, Vol. 19, No. 2, pp. 187-207.

Tonsor, G. and A. Featherstone (2009), "Production efficiency of specialized swine producers", *Review of Agricultural Economics*, Vol. 31, No. 3, pp. 493-510.

Van Berkum, S. (2009), *An Assessment of the Competitiveness of the Dairy Supply Chain in New Member States, Candidate Countries and Potential Candidate Countries*, final report, AgriPolicy, May.

Van Duren, E., L. Martin, and R. Westgren (1991), "Assessing the competitiveness of Canada's agrifood industry", *Canadian Journal of Agricultural Economics*, Vol. 39, pp. 727-738.

Venturini, L. and S. Boccaletti (1998), "Sophisticated consumers and export success, but problems in the home retail sector: The Italian pasta industry", in Traill, B., and E. Pitts (eds), *Competitiveness in the Food Industry*, Blackie Academic & Professional, London, Chapter 6, pp. 179-208.

Viaene, J. and X. Gellynck (1998), "Small firms, old traditions equals low profit: Pigmeat processing in Belgium", in Traill, B., and E. Pitts (eds), *Competitiveness in the Food Industry*, Blackie Academic and Professional, London, Chapter 5, pp. 149-178.

Weersink, A., C. Turney, and A. Godah (1990), "Decomposition measures for technical efficiency for Ontario dairy farms", *Canadian Journal of Agricultural Economics*, Vol. 38, No. 3, pp. 439-456.

Wijnands, J., H. Bremmers, B. van der Meulen, and K. Poppe (2008), "An economic and legal assessment of the EU food industry's competitiveness", *Agribusiness*, Vol. 24, No. 4, pp. 417-439.

World Bank (2007), *World Development Report 2008: Agriculture for Development*, Washington D.C., World Bank, October. Available at: *siteresources.worldbank.org/INTWDR2008/Resources/WDR_00_book.pdf*

Yee, J., M. Ahearn, and W. Huffman (2004), "Links among farm productivity, off-farm work, and farm size in the Southeast", *Journal of Agricultural and Applied Economics*, Vol. 36, No. 3, pp. 591-603.

Zhengfei, G. and A. Oude Lansink (2006), "The source of productivity growth in Dutch agriculture: A perspective from finance", *American Journal of Agricultural Economics*, Vol. 88, No. 3, pp. 644-656.

Zhu, X., R. Demeter, and A. Oude Lansink (2008a), *Competitiveness of Dairy Farms in Three Countries: The Role of CAP Subsidies*, paper presented at the 12th EAAE Congress, Gent, Belgium, 27-30 August.

Zhu, X., G. Karagiannis, and A. Oude Lansink (2008b), *Analyzing the Impact of Direct Subsidies on the Performance of the Greek Olive Farms with a Non-Monotonic Efficiency Effects Model*, paper presented at the 12th EAAE Congress, Gent, Belgium, 27-30 August.

Annex A.

Summary of evidence on productivity and competitiveness in the agri-food sector

Table A.1. Competitiveness and profitability of the food processing sector

Study	Countries	Sector	Period	Indicators	Results
Bavorova (2003)	Czech Republic	Sugar industry	1989 and 1999	Production costs Labour productivity (value-added per employee)	Higher concentration results in economies of scale and lower costs of production. Labour productivity increased six-fold. Sugar yields increase more than in the EU15 between 1989 and 2000 but remain lower in 2000.
Buccola *et al.* (2000)	United States	Grain processing industry	1958-94	TFP growth	Steady growth over the period except for the baking industry where productivity declined.
Chan-Kang *et al.* (1999)	United States and Canada	Food manufacturing sector	1963-92	TFP growth	Canada lagging behind during the period.
Gopinath (2003)	13 OECD countries	Food processing industry	1975-95	TFP growth	The United States had the highest TFP but Denmark had the highest TFP growth
Ruan and Gopinath (2008)	34 developed and developing countries	Five food processing industries	1993-2000	TFP growth	Highest growth rate in the United States; Increase for meat, fish and dairy processing industries; decrease for oils and fats.
Fischer and Schornberg (2007)	13 EU member states	Ten products of the food and drink manufacturing sector	1995-98 to 1999-2002	Index comprising multiple indicators of competitiveness and productivity	The European Union competitiveness increased slightly between the two periods and the index has converged across countries.
Van Duren *et al.* (1991)	Canada, European Union, United States	Agri-food sector	1986	- Ratio value added to sales - Value added per worker - Value added per plant	The United States more competitive than Canada more competitive than the European Union; Canada most competitive for meats; the European Union and the United States highly competitive for beverages.
Viaene and Gellynck (1998)	Belgium	Pigmeat processing sector	1987-93	- Net profit relative to sales - Sales divided by business assets - Net profit to own funds - Financial leverage	Poor profitability
Wijnands *et al.* (2008)	EU15 *vis-à-vis* Australia, Brazil, Canada and the United States	Food industry	1996-2004	Growth in RCA and EMS Labour productivity	Brazil has the lowest score and the United States the highest. EU15 labour productivity is about average.

Most studies that report trade measures of competitiveness consider generally trade in agricultural and agro-food products.

Source: Latruffe (2010).

Table A.2. Total factor productivity in agriculture

Study	Countries	Sector	Period	Indicators	Results
Ball *et al.* (2006)	United States, 11 EU countries	Agricultural outputs and inputs	1973-2002	TFP	TFP increased consistently for all countries in most years. Rapid growth in Spain and Sweden, from a lower base. US TFP grew faster than in most EU countries and is the highest in 2002.
Ball *et al.* (2001)	United States, 9 EU countries	Agricultural outputs and inputs	1973-1993	Eltetö-Köves-Szulc TFP indices	Largest gains in France. Belgium and the Netherlands have highest TFP, Ireland is behind; convergence between country.
Ball *et al.* (1997)	United States	Agriculture	1948-94	Fisher TFP indices	Annual growth rate of 1.94%, lowest rates between 1948 and 1957; highest rates in 1966-69 and 1989-94.
Brümmer *et al.* (2002)	Poland, Germany, Netherlands	Dairy farms	1991-94	Technical efficiency; Malmquist TFP	Deterioration in Poland (-5%), increase in Germany (6%) and Netherlands (3%).
Carroll *et al.* (2009)	Ireland	Cattle, cereals, dairy, sheep farms	1996-2006	Technical efficiency; TFP growth	Average growth is highest for cattle rearing (2%) and dairy (1.4%), low for sheep, negative for cereals.
Davidova et al. (2003)	Czech Republic	Agriculture	1998-99	Cost/revenues; Tornquist indices	40% of farms were productive (index > 1)
Fogarasi and Latruffe (2009)	France and Hungary	Cereals, oilseeds and protein crops and dairy	2001-04	Malmquist TFP	No change with technical efficiency increase being offset by slight technological deterioration, or deterioration (Hungarian dairy farms)
Galonopoulos *et al.* (2008)	32 EU and Mediterranean countries	Agriculture	1966-2002	Malmquist TFP	High productivity in EU15 and Central and Eastern Europe; Low productivity in Southern countries Convergence from 1990
Hadley (2006)	England and Wales	Eight farm types	1982-2002	Malmquist TFP	Positive technological change
Latruffe *et al.* (2008)	Poland	All farms	1996-2000	Technological change; Malmquist TFP	Deterioration (-2% on average)

TFP: Total Factor Productivity.

Source: Latruffe (2010).

Table A.3. Components of productivity in agriculture

Study	Countries	Sector	Period	Indicators	Results
Brümmer *et al.* (2002)	Poland, Germany, Netherlands	Dairy farms	1991-94	Technical efficiency; Malmquist TFP	Higher in Poland, then Netherlands, than Germany
Carroll *et al.* (2009)	Ireland	Cattle, cereals, dairy, sheep	1996-2006	Technical efficiency change; TFP	Deterioration followed by progress; strongest progress in 2005/06.
Hadley (2006)	England and Wales	Eight farm types	1982-2002	Technical efficiency + change	High scores Change: zero or negative
Hadley (2006)	England and Wales	Eight farm types	1982-2002	Technological change	Positive in all farm types; strongest progress by cereal and mixed farms, smallest for poultry farms.
Giannakas *et al.* (1998)	Saskatchewan	Crop farms	1987-1995	Technical efficiency	Increasing trend
Latruffe *et al.* (2005)	Poland	Specialised Crop and livestock farms	1996 and 2000	Technical efficiency	Decrease
Latruffe *et al.* (2008)	Poland	All farms	1996-2000	Technological change	Deterioration (-6% on average)
Nasr *et al.* (1998)	Illinois	Grain farms	1988-94	Technical efficiency	Increasing trend
Zhu *et al.* (2008a)	Germany, Netherlands, Sweden	Dairy farms	1995-2004	Technical efficiency change	Increase in Germany (1%) and the Netherlands (2.8%); decrease in Sweden (-1.1%).

Source: Latruffe (2010).

Table A.4. Partial factor productivity in agriculture

Study	Countries	Sector	Period	Indicators	Results
Alston *et al.* (2008a)	Developing and developed countries	Crops	1961-2006	Yields	Maize rice and wheat yield growth slows down after 1990. Maize yield growth is lower in developing countries than in developed countries after 1990.
Alston *et al.* (2008b)	United States	Crops	1961-2006	Land and labour productivity	Faster growth before 1990.
Bureau and Butault (1992)	EU member states	Wheat, sugar beet, hog and milk	1984	Costs of production Partial productivity indices	On the basis of labour productivity, the United Kingdom and France most competitive for wheat production, France for sugar beet, Netherlands for hogs and Belgium, Ireland, the Netherlands and the United Kingdom for milk
Mulder *et al.* (2004)	Mercosur-European Union	Several protected products	1995	Unit labour costs All input costs Labour and land productivity (value of output/input)	Labour and land productivity much lower in Mercosur countries than in EU15 (but costs of production are lower)
Thorne (2005)	Denmark, Germany, France, Ireland, Italy, United Kingdom	Cereal production	1996-2000	Cost indicators; yields; Labour productivity	Italy lags behind of studied countries
Van Berkum (2009)	12 new EU member states 8 candidates	Dairy sector	2006	Gross margin as a % of revenue; milk yield	Cow milk yield close to EU15 average only in the Czech Republic, Hungary and Estonia, lowest in the Balkans

Source: Latruffe (2010).

Table A.5. Competitiveness of agriculture using trade measures

Study	Countries	Sector	Period	Indicators	Results
Ball *et al.* (2006)	United States, 11 EU member states	Agricultural outputs and inputs	1973-2002	PPP prices	US agriculture is more competitive as input prices are lower; lowest competitiveness found in Ireland, Italy, Portugal and Spain.
Banterle and Carraresi (2007)	EU member states	Prepared swine meat sector	2000-2003	EMS, TCA, RMA, clusters based on all indicators	Italy highest EMS, followed by Germany; Denmark highest RCA score, then Italy; Low RMA in Finland, Italy and Spain; Highest growth in RMA and EMS in Austria.
Bavorova (2003)	Czech Republic	Sugar industry	1988-99	RXA, RMA and RTA	RXA indicate international competitive disadvantage every year, while RMA and RTA reveal a competitive advantage in 1994-98.
Bojnec and Fertő (2009)	Eight Central and Eastern European and Balkan countries	Agri-food sector (four categories of products)	1995-2007	RXA, RMA and RTA	Export competitive advantage is highest for raw and intermediate processed products in all countries
Carraresi and Banterle (2008)	Several EU members	Agro-food and agricultural sectors	1991-2006	RCA, RXA, RMA, EMS, NEI, clusters	Cluster analysis identifies 3 groups of countries
Drescher and Maurer (1999)	Germany compared to other EU members	Dairy products	1983-1993	EMS, RCA	Germany had one of the most competitive dairy sector in the EU, in particular in milk and evaporated milk products
Fertő and Hubbard (2003)	Hungary	Agro-food sector (22 products)	1992-98	RXA, RMA, RTA and RC	Revealed competitive advantage for half of the products, in particular cereals, meat, sugar and live animals; but it decreases over the period.
Mulder *et al.* (2004)	EU and Mercosur countries	Agriculture and agro-food sector	1991-99	RER and relative exchange rates	Competitiveness vis-à-vis the European Union decreases in Mercosur countries (except Paraguay) until 1998; in 1999, the devaluation of the Brazilian currency boosts competitiveness.
Qineti *et al.* (2009)	Slovak Rep. and EU27 *vis-à-vis* Russia and Ukraine	Agri-food sectors	2002-06	RCA growth	Since EU enlargement in 2004, the number of EU commodities with a comparative advantage over Russia and Ukraine has declined. Slovakia has loss some competitive advantage over Russia but gained over Ukraine.
Toming (2007)	Estonia	Agro-food industry	1999-2005	Value of exports to other EU members	Competitiveness has increased since accession

Table A.5. Competitiveness of agriculture using trade measures (*cont.*)

Study	Countries	Sector	Period	Indicators	Results
Van Berkum (2009)	12 new EU member states and 8 candidates	Dairy sector	2006	Trade position and trend	Most countries were net exporters and the Baltic countries and Poland had increased their surpluses since the 1990s.
Venturini and Boccaletti (1998)	Italy versus other EU members	Pasta processing	1988-92	RCA	High and increasing competitiveness for Italy
Wijnands *et al.* (2008)	EU15 *vis-à-vis* Australia, Brazil, Canada and the United States	Food industry	1996-2004	Growth in RCA and EMS Labour productivity	EU15 has a low competitiveness compared to Brazil, but higher than in the united States for EMS (lower for RCA growth)

Notes

RER: Real Exchange Rate, i.e. the ratio of the price index of tradable commodities and the price of non-tradable ones; PPP: Purchasing Power Parity;

RCA: Revealed Comparative Advantage, i.e. the ratio of the share of trade in commodity i over all commodities in a country and the same share in all other countries;

Relative Import Advantage (RMA) and Relative Export Advantage (RXA) are the same for imports and exports respectively; The Relative Trade Advantage (RTA) is the difference between RXA and RMA; Revealed competitiveness (RC) is the difference between the logarithms of RXA and RMA;

NEI: Net Export Index is the difference between exports and imports divided by the total value of trade (imports plus exports); EMS: Export Market Shares; TCA: Trade Competitive Advantage.

Source: Latruffe (2010).

Table A.6. Competitiveness of agriculture using costs measures: Domestic resource cost

Study	Countries	Sector	Period	Indicators	Results
Banse *et al.* (1999)	Hungary	Various crop and livestock sectors	1990-96	DRC	Livestock sector less and less competitive Wheat competitive
Bojnec (2003)	Central and Eastern European Countries (CEECs)	Various crop and livestock sectors	1989-98	DRC	Livestock production was less internationally competitive than crop production.
Gorton *et al.* (2000)	Bulgaria, Czech Rep. compared to EU15 and world	Main commodities	1994-96	DRC	High competitiveness of wheat and barley both *vis-à-vis* EU15 and world. Competitive in milk and beef relative to the European Union but not world.
Gorton *et al.* (2001)	Poland	Eight commodities	1996 and 1998	DRC at farm level	Crops more internationally competitive than livestock. Competitiveness worsens
Gorton and Davidova (2001)	CEECs		1992 and 1998	DRC at farm level	Crops most competitive. Higher competitiveness in CEECs than in the European Union
Liefert (2002)	Russia	Several output and inputs	1996-97	SCB ratios	Less competitive in meat than in crops. More competitive in outputs than in inputs (except natural gas)
Nivievskyi and von Cramon-Taubadel (2008)	Ukraine	Dairy production	2004-05	DRC and SCB at farm level	15% of farms were competitive in 2005 (19% in 2004)

Notes

DRC: Domestic Resource Cost, which compares the opportunity costs of domestic production with the value added it generates.

SCB: Social Cost-Benefit, which is the ration of the sum of domestic (non-tradable) and tradable input costs to the price of the good considered.

Source: Latruffe (2010).

Table A.7. Competitiveness of agriculture using costs measures: Costs of production

Study	Countries	Sector	Period	Indicators	Results
Ahearn *et al.* (1990)	United States and Canada	Wheat	1986-87	Costs of production	Higher in the United States than in Canada
Bureau and Butault (1992)	EU member states	Wheat, sugar beet, hog and milk	1984	Costs of production Partial productivity indices	On the basis of costs of production, United Kingdom and France most competitive for wheat production, Belgium and France for sugar beet, Ireland, the Netherlands and the United Kingdom for hog and Greece for milk
Bureau *et al.* (1992)	EU MS and the United States	Wheat	Average 1984-86	Costs of production	The United States has by far lowest costs, Italy highest
Thorne (2005)	Denmark, Germany, France, Ireland, Italy, United Kingdom	Cereal production	1996-2000	Cost indicators; Yields, labour productivity	Depends if family labour and assets are included or not
Mulder *et al.* (2004)	Brazil-European Union Mercosur-EU	Several protected products	1995	Unit labour costs All input costs Labour and land productivity	Brazil costs are 15.5% of EU costs and 5% of French costs Mercosur more competitive for all products except bananas

Source: Latruffe (2010).

Table A.8. Competitiveness of agriculture using profitability measures

Study	Countries	Sector	Period	Indicators	Results
Davidova *et al.* (2003)	Czech Rep.	Agriculture	1998-99	Cost/revenues; Tornquist indices	Most farms not profitable, even when family inputs not considered
Van Berkum (2009)	12 new EU MS 8 candidates	Dairy sector	2006	Gross margin as a % of revenue; milk yield	62% for the EU15, only Slovenia, Bosnia and Poland have a higher ratio

Source: Latruffe (2010).

Table A.9. Competitiveness and farm size

Study	Countries	Sector	Period
Larger farms are better performers			
Weersink *et al.* (1990)	Ontario, Canada	Dairy farms	1987
Hallam and Machado (1996)	Portugal	Dairy farms	1989-92
Nasr *et al.* (1998)	Illinois, United States	Grain farms	1988-94
Sharma *et al.* (1999)	Hawai	Pig farms	1994
Brümmer and Loy (2000)	Germany	Dairy farms	1987-94
Huffman and Evenson (2001)	United States	Livestock farms	1953-82
Yee *et al.* (2004)	United States	Agriculture	1960-96
Latruffe *et al.* (2004 and 2008)	Poland	Crop farms	1996-2000
Hadley (2006)	England and Wales	Various farm types	1982-2002
Rios and Shively (2006	Vietnam	Coffee farms	2004
Emvalomatis *et al.* (2008)	Greece	Cotton farms	1996-2000
Zhu *et al.* (2008a)	Germany and Sweden	Dairy farms	1995-2004
Carrol *et al.* (2009)	Ireland	Several livestock farm types	1996-2006
Smaller farms are better performers			
Munroe (2001)	Poland	Agriculture	1996
Huffman and Evenson (2001)	United States	Crop farms	1953-82
O'Neill and Matthews (2001)	Ireland	Agriculture	1984-98
Zhu *et al.* (2008b)	Greece	Olive farms	1995-2004
U-shaped relationship			
Helfand and Levine (2004)	Brazil	Agriculture	1995
Latruffe *et al.* (2005)	Poland	Livestock farms	1996-2000
Tonsor and Feathernstone (2009)	United States	Pig farms	2004
Depends on the farm size variable			
Bojnec and Latruffe (2007)	Slovenia	Agriculture	1994-2003

Source: Latruffe (2010).

Table A.10. Farm competitiveness and support

Study	Countries	Sector	Period	Indicators
Positive correlation				
Bezlepkina *et al.* (2005)	Russia	Dairy farms	1995-2001	Profit and subsidies
Rezitis *et al.* (2003)	EU countries	Agriculture		Farm efficiency and farm subsidies
Emvalomatis *et al.* (2008)	Greece	Cotton farms	1996-2000	Farm efficiency and share of support in farm income
Giakannas *et al.* (2001)	Saskatchewan	Crop farms	1987-995	Farm efficiency and share of support in output or gross margin
Hadley (2006)	England and Wales	Different farm types	1982-2002	Idem
Zhu *et al.* (2008a)	Germany, Netherlands, Sweden	Dairy farms	1995-2004	Idem
Zhu *et al.* (2008b)	Greece	Olive farms	1995-2004	Idem
Bojnec and Latruffe (2009)	Slovenia	Agriculture		Idem
Fogarasi and Latruffe (2009)	France and Hungary	Crop and dairy	2001-04	Idem
Latruffe *et al.* (2009)	France	Crop and beef farms	2000	Idem
Bakucs *et al.* (2010)	Hungary	All farms	2001-05	Idem
Hadley (2006)	England and Wales	Dairy and beef farms	1982-2002	Technical efficiency
Huffman and Evenson (2001)	US states	Crop and livestock sectors	1953-82	Price support and TFP
Negative correlation				
Banse *et al.* (1999)	Hungary	Various crop and livestock sectors	1992-96	DRC and PSE
Nivievskyi and von Cramon-Taubadel (2008)	Ukraine	Dairy production	2004-05	SCB and subsidies
Giannakas *et al.* (2001); Rezitis *et al.* (2003); Emvalomatis *et al.* (2008); Zhu *et al.* (2008a); Zhu *et al.* (2008b); Bojnec and Latruffe (2009);, Fogarasi and Latruffe (2009); Latruffe *et al.* (2009); Bakucs *et al.* (2010)	See above	See above	See above	Support and technical efficiency
Hadley (2006)	England and Wales	Cereal, sheep and general cropping and mixed farms	1982-2002	Support and technical efficiency
Lachaal (1994)	United States	Dairy sector	1972-92	Government expenditures and technical efficiency
Non-significant correlation				
Peterson and Valluru (2000)	40 countries	Agriculture	1992	Trade and PSE
Yee *et al.* (2004)	US States	agriculture	1960-96	Commodity payments and TFP
Makki *et al.* (1999)	United States	Agriculture	1930-90	Commodity payments and TFP

Source: Latruffe (2010).

ORGANISATION FOR ECONOMIC CO-OPERATION AND DEVELOPMENT

The OECD is a unique forum where governments work together to address the economic, social and environmental challenges of globalisation. The OECD is also at the forefront of efforts to understand and to help governments respond to new developments and concerns, such as corporate governance, the information economy and the challenges of an ageing population. The Organisation provides a setting where governments can compare policy experiences, seek answers to common problems, identify good practice and work to co-ordinate domestic and international policies.

The OECD member countries are: Australia, Austria, Belgium, Canada, Chile, the Czech Republic, Denmark, Estonia, Finland, France, Germany, Greece, Hungary, Iceland, Ireland, Israel, Italy, Japan, Korea, Luxembourg, Mexico, the Netherlands, New Zealand, Norway, Poland, Portugal, the Slovak Republic, Slovenia, Spain, Sweden, Switzerland, Turkey, the United Kingdom and the United States. The European Union takes part in the work of the OECD.

OECD Publishing disseminates widely the results of the Organisation's statistics gathering and research on economic, social and environmental issues, as well as the conventions, guidelines and standards agreed by its members.

OECD PUBLISHING, 2, rue André-Pascal, 75775 PARIS CEDEX 16
(51 2011 13 1 P) ISBN 978-92-64-16680-6 – No. 59713 2011-01